A Second Step to Mathematical Olympiad Problems

Mathematical Olympiad Series

ISSN: 1793-8570

Series Editors: Lee Peng Yee *(Nanyang Technological University, Singapore)*
Xiong Bin *(East China Normal University, China)*

Published

Vol. 1   A First Step to Mathematical Olympiad Problems
by Derek Holton (University of Otago, New Zealand)

Vol. 2   Problems of Number Theory in Mathematical Competitions
by Yu Hong-Bing (Suzhou University, China)
translated by Lin Lei (East China Normal University, China)

Vol. 3   Graph Theory
by Xiong Bin (East China Normal University, China) &
Zheng Zhongyi (High School Attached to Fudan University, China)
translated by Liu Ruifang, Zhai Mingqing & Lin Yuanqing
(East China Normal University, China)

Vol. 4   Combinatorial Problems in Mathematical Competitions
by Yao Zhang (Hunan Normal University, P. R. China)

Vol. 5   Selected Problems of the Vietnamese Olympiad (1962–2009)
by Le Hai Chau (Ministry of Education and Training, Vietnam)
& Le Hai Khoi (Nanyang Technology University, Singapore)

Vol. 6   Lecture Notes on Mathematical Olympiad Courses:
For Junior Section (In 2 Volumes)
by Xu Jiagu

Vol. 7   A Second Step to Mathematical Olympiad Problems
by Derek Holton (University of Otago, New Zealand &
University of Melbourne, Australia)

Derek Holton

University of Otago, New Zealand & University of Melbourne, Australia

Vol. 7 | Mathematical Olympiad Series

A Second Step to Mathematical Olympiad Problems

World Scientific

NEW JERSEY · LONDON · SINGAPORE · BEIJING · SHANGHAI · HONG KONG · TAIPEI · CHENNAI

Published by

World Scientific Publishing Co. Pte. Ltd.

5 Toh Tuck Link, Singapore 596224

USA office: 27 Warren Street, Suite 401-402, Hackensack, NJ 07601

UK office: 57 Shelton Street, Covent Garden, London WC2H 9HE

British Library Cataloguing-in-Publication Data
A catalogue record for this book is available from the British Library.

Mathematical Olympiad Series — Vol. 7
A SECOND STEP TO MATHEMATICAL OLYMPIAD PROBLEMS

Copyright © 2011 by World Scientific Publishing Co. Pte. Ltd.

ISBN-13 978-981-4327-87-9 (pbk)
ISBN-10 981-4327-87-5 (pbk)

Typeset by Stallion Press
Email: enquiries@stallionpress.com

Printed in Singapore.

To Lee Peng Yee, a true friend and all-round nice guy, who was instrumental in finding a publisher for both Steps.

Foreword

This is the second of two volumes on some higher level problem solving and largely contains material that is not taught in most Western schools. Like the first volume (*A First Step to Mathematical Olympiad Problems*), most of the content here was developed initially as an aid for New Zealand students who were hoping to get to an International Mathematical Olympiad (IMO). This original material was presented in the form of 15 yellow booklets. The first eight of these are the basis of *First Step*; the remaining seven underpin this volume. However, it should be noted that there are also new pieces here. The whole of Chapter 3 was not in the initial yellow booklets neither were many sections and problems that are dotted around the rest of the book.

This book should help improve the background of students preparing for regional and national mathematics competitions by providing them with theorems, ideas and techniques that they may not have already met. It will also give them suggestions on how to approach problems that they have not seen before. But the step to the IMO from regional and national competitions is likely to be a big one and the First and Second Steps should not be considered as a total preparation for an IMO. Hopefully though, they are good stepping stones along the way.

But I hope that the book will also be read by both students and teachers who are interested in mathematics. There are sufficiently many problems here to while away many happy and frustrating hours. I hope that this book encourages many readers to simply enjoy mathematics for its own sake and for the pleasure of getting a problem out.

Finally I would like to thank Irene Goodwin (who is now, unfortunately, no longer with us), Lenette Grant and Leanne Kirk for their help to me while the yellow booklets were being prepared and in the many years since, and to Zhang Ji for her work in turning yellow booklets into Steps.

Contents

Chapter 1

Combinatorics

After a quick revision of some ideas discussed in the earlier book, *A First Step to Mathematical Olympiad Problems*, I'll consider partial fractions, recurrence relations, generating functions, Fibonacci numbers and derangements. The net effect should be to increase your ability to count certain sets of objects that have a well-defined rule.

1.1. A Quick Reminder

In Chapter 4 of *First Step*,[a] I mentioned Arithmetic Progressions (A.P's). It's been preying on my mind that I omitted to mention Gauss then. Young Carl Friedrich was brought up in Braunschweig (Brunswick) (look that up on Google Earth) which is now in the state of Lower Saxony in the Federal Republic of Germany. (Braunschweig was actually the site of the 30th IMO in 1989.) The story goes that one day one of his schoolteachers asked the class to find the sum of the numbers from 1 to 100. Now it's clear that the teacher had intended this to keep the students quiet for the hour so that he could get on with something more important. You can imagine the teacher's chagrin when young Carl almost immediately reported a total of 5050. Can you do this inside two minutes? Think about it before you read on.

Of course, like anything else, it's very simple when you know how. What Gauss had noted was that $1+100 = 101$, that $2+99 = 101$, that $3+98 = 101$ and so on. If you call the sum of the first 100 whole numbers, S, you can see that we can write S in two ways — forwards and backwards. We've done

---

[a] *A First Step to Mathematical Olympiad Problems*, published by World Scientific, 2009.

it below.

$$S = 1 + 2 + 3 + 4 + \cdots + 100,$$

$$S = 100 + 99 + 98 + 97 + \cdots + 1.$$

Adding these two rows we get

$$2S = 101 + 101 + 101 + 101 + \cdots + 101.$$

$$\therefore \quad 2S = 100 \times 101.$$

And so $S = 50 \times 101 = 5050$.

In Chapter 4 of *First Step*, you can see that this works for **any** Arithmetic Progression, AP for short, that is any sequence of terms where the difference between consecutive terms is a constant d. So you should now know how to find S_n below:

$$S_n = a + (a + d) + (a + 2d) + \cdots + [a + (n - 1)d],$$

where a and d are fixed and n is the number of terms being added together.

Exercises

1. Find out when Gauss was born, when he died, what he did in Göttingen, how far it is from Göttingen to Braunschweig and what was the importance of the 17 sided figure he constructed.
2. Find the sum of the first 23 terms of the following APs.
   (i) 1, 5, 9,...;   (ii) 2, 12, 22,...;   (iii) 3, 10, 17,...;
   (iv) 1/2, 3/2, 5/2,...;   (v) 4, 2, 0,...;   (vi) 26, 22, 18,....
3. Find the value of the following sums.
   (i) $1 + 5 + 9 + \cdots + 45$;   (ii) $2 + 12 + 22 + \cdots + 102$;
   (iii) $3 + 10 + 17 + \cdots + 73$;   (iv) $1/2 + 3/2 + 5/2 + \cdots + 99/2$;
   (v) $4 + 2 + 0 + \cdots + (-32)$;   (vi) $26 + 22 + 18 + \cdots + (-2)$;
4. What is the significance of "the average of the first and the last multiplied by the number of terms"?

Chapter 4 of *First Step* also introduced the Euclidean Algorithm. This gives you a way of finding the highest common factor of two numbers and that leads on to the solution of certain Diophantine equations of the form $as + bt = r$.

You should remember (if you don't, it's probably a good idea to go and check it out), that to find the highest factor of 78 and 30 we adopt the following approach

$$78 = 2 \times 30 + 18,$$

$$30 = 1 \times 18 + 12,$$

$$18 = 1 \times 12 + 6,$$

$$12 = 2 \times 6.$$

Since 6 is the last non-zero remainder in this process, then $6 = (78, 30)$. That is, 6 is the highest common factor of 78 and 30. But working back from there we see that

$$
\begin{aligned}
6 &= 18 - 1 \times 12 \\
&= 18 - (30 - 1 \times 18) \\
&= 2 \times 18 - 30 \\
&= 2 \times (78 - 2 \times 30) - 30 \\
&= 2 \times 78 - 5 \times 30.
\end{aligned}
$$

In other words, if we want to solve $78a + 30b = 6$, one solution is $a = 2$, $b = -5$. But Chapter 4 of *First Step* showed that there are in fact an infinite number of solutions of $78a + 30b = 6$. It also showed how to find all these solutions.

Exercises

5. Using the Euclidean Algorithm, find the highest common factor of the following pairs of numbers.
   (i) 78, 45; (ii) 121, 33; (iii) 151, 72.
6. Use the last exercise to find all solutions of the following equations.
   (i) $78a + 45b = 3$; (ii) $121a + 33b = 11$; (iii) $151a + 72b = 1$.

The other thing in Chapter 4 of *First Step* that I wanted to mention here is the technique that enabled us to find things like $\sum_{r=1}^{n} r^2$.

For instance, suppose we want to find a simple expression for $\sum_{r=1}^{n} r^6$. Then we work as follows.

$$(n+1)^7 - n^7 = 7n^6 + 21n^5 + 35n^4 + 35n^3 + 21n^2 + 7n + 1,$$
$$n^7 - (n-1)^7 = 7(n-1)^6 + 21(n-1)^5 + 35(n-1)^4 + 35(n-1)^3$$
$$+ 21(n-1)^2 + 7(n-1) + 1,$$
$$(n-1)^7 - (n-2)^7 = 7(n-2)^6 + 21(n-2)^5 + 35(n-2)^4 + 35(n-2)^3$$
$$+ 21(n-2)^2 + 7(n-2) + 1,$$

$$\vdots$$

$$3^7 - 2^7 = 7 \times 2^6 + 21 \times 2^5 + 35 \times 2^4 + 35 \times 2^3$$
$$+ 21 \times 2^2 + 7 \times 2 + 1,$$
$$2^7 - 1^7 = 7 \times 1^6 + 21 \times 1^5 + 35 \times 1^4 + 35 \times 1^3$$
$$+ 21 \times 2^2 + 7 \times 1 + 1.$$

If we add all these equations up, then we get

$$(n+1)^7 - 1^7 = 7\sum_{r=1}^{n} r^6 + 21\sum_{r=1}^{n} r^5 + 35\sum_{r=1}^{n} r^4 + 35\sum_{r=1}^{n} r^3$$
$$+ 21\sum_{r=1}^{n} r^2 + 7\sum_{r=1}^{n} r + \sum_{r=1}^{n} 1.$$

Consequently if we knew $\sum_{r=1}^{n} r^5, \sum_{r=1}^{n} r^4, \sum_{r=1}^{n} r^3, \sum_{r=1}^{n} r^2, \sum_{r=1}^{n} r$ and $\sum_{r=1}^{n} 1$, then we could solve to find $\sum_{r=1}^{n} r^6$.

Exercise

7. Find the following sums in terms of n,
   (i) $\sum_{r=1}^{n} 1$;   (ii) $\sum_{r=1}^{n} r$;   (iii) $\sum_{r=1}^{n} r^3$;   (iv) $\sum_{r=1}^{n} r^4$.

In the next section we will show that there are other uses for the method of adding differences that we have just been using to find $\sum_{r=1}^{n} r^k$, where k is a fixed integer.

1.2. Partial Fraction

It should be no problem for you to see that $\frac{1}{3} - \frac{1}{4} = \frac{1}{12}$ and $\frac{1}{1-x} + \frac{1}{1+x} = \frac{2}{1-x^2}$. In that case it's obvious that

$$\frac{1}{12} = \frac{1}{3} - \frac{1}{4} \quad \text{and} \quad \frac{2}{1-x^2} = \frac{1}{1-x} + \frac{1}{1+x}.$$

What's the big deal here? Well sometimes it's useful to break a fraction down into fractional parts or **Partial Fractions**. Hence the partial fractions of $\frac{1}{12}$ are $\frac{1}{3}$ and $-\frac{1}{4}$ and if you add the partial fractions together you get the original fraction. Similarly the partial fractions of $\frac{2}{1-x^2}$ are $\frac{1}{1-x}$ and $\frac{1}{1+x}$.

The problem I want to look at now, is first to find partial fractions for any fraction and second to find some use for them.

So how do we find partial fractions? Well, it's really not too difficult. Take $\frac{1}{12}$ as an example. We know that $12 = 3 \times 4$. So assume that

$$\frac{1}{12} = \frac{a}{3} + \frac{b}{4}$$

and we'll try to find a and b. To do this, put both sides over a common denominator. So

$$\frac{1}{12} = \frac{4a + 3b}{12}.$$

All we have to do now is to solve the equation $4a + 3b = 1$. This should present no problem because we considered this idea in Section 1.1 above.

Now

$$4 = 3 + 1.$$

$$3 = 1 \times 3.$$

So $1 = 4 - 3$.

I suppose you could have guessed that. One solution of $4a + 3b = 1$ then, is $a = 1, b = -1$. Consequently $\frac{1}{12} = \frac{1}{3} - \frac{1}{4}$. We know this already but now we have a way of getting it from $\frac{1}{12}$ *and* we also know there must be lots of other ways of writing $\frac{1}{12}$.

For instance, $4a + 3b = 1$ also has the solution $b = 3, a = -2$. We can therefore write

$$\frac{1}{12} = \frac{3}{4} - \frac{2}{3}.$$

From what I said in Section 1.1, it ought to be clear that there are an infinite number of ways of finding partial fractions for $\frac{1}{12}$ in terms of thirds and quarters.

Exercises

8. Find several ways of writing $\frac{1}{14}$ in terms of the fractions $\frac{1}{7}$ and $\frac{1}{2}$.

9. Find a couple of partial fraction decompositions for the following.

(i) $\dfrac{1}{10}$; (ii) $\dfrac{1}{18}$; (iii) $\dfrac{1}{15}$;

(iv) $\dfrac{2}{63}$; (v) $\dfrac{3}{56}$; (vi) $\dfrac{4}{35}$.

10. Use partial fractions to find the sum of the series

$$\frac{1}{12} + \frac{1}{20} + \frac{1}{30} + \frac{1}{42} + \cdots + \frac{1}{9900}.$$

11. Show that $\sum_{r=3}^{\infty} \frac{1}{r(r+1)} = \frac{1}{3}$.

So we know how to get partial fractions for regular fractions. How can we handle fractions with x's in them? Take $\frac{2}{1-x^2}$ for instance.

Now because $1 - x^2 = (1-x)(1+x)$, it seems reasonable to look for a and b such that

$$\frac{2}{1-x^2} = \frac{a}{1-x} + \frac{b}{1+x}.$$

In the $\frac{1}{12}$ example, we next put everything over the common factor of 12. So let's repeat that here with $1 - x^2$.

So $\frac{2}{1-x^2} = \frac{a(1+x)+b(1-x)}{1-x^2}$ and we have to find a and b to satisfy

$$2 = a(1+x) + b(1-x) \cdots . \tag{1}$$

There is more than one way to do this. I'll try to show you what I think is the **quickest** way. Now equation (1) is true for all values of x. So if we can substitute some values of x which make it easy to find a and b, then we can do that. Surely the best values to substitute then, are $x = 1$ and $x = -1$.

$$x = 1: \quad 2 = a(1+1) + b(1-1) \quad \therefore \ a = 1.$$
$$x = -1: \quad 2 = a(1-1) + b(1+1) \quad \therefore \ b = 2.$$

So

$$\frac{2}{1-x^2} = \frac{1}{1-x} + \frac{1}{1+x}.$$

This method of solving equation (1) by choosing x so that factors disappear, will work every time.

Exercises

12. Find partial fraction decompositions for the following fractions.

(i) $\dfrac{4}{4-x^2}$;   (ii) $\dfrac{9}{9-x^2}$;   (iii) $\dfrac{16}{16-x^2}$;

(iv) $\dfrac{1}{(x-1)(x+2)}$;   (v) $\dfrac{1}{(x+1)(x+2)}$;   (vi) $\dfrac{1}{x(x-1)}$.

13. Find the partial fraction decomposition for
$$\frac{1}{x(x-1)(x+1)}.$$

14. Are the partial fraction decompositions of Exercises 12 and 13 unique? In other words, are there other families of solutions, as there were for fractions of the type $\frac{1}{12}$ or is there just one solution each time?

Actually, the method I showed you earlier is not the quickest. I lied! Apologies. (This just shows that you can never trust a mathematician. You should check everything that I say. If nothing else, I do make errors from time to time — even in books.) It turns out that **the cover up method** is quicker. However, this is a deep dark secret, reserved for the most sophisticated mathematical gurus. It is with some hesitation that I reveal it now to you. I sincerely expect great retribution will be heaped on my head. Undoubtedly I will be struck off all sophisticated mathematical gurus' Christmas cards' lists.

Now take $\frac{2}{1-x^2} = \frac{1}{(1-x)(1+x)}$.

Remember I wanted to find a and b in $\frac{a}{1-x} + \frac{b}{1+x}$.

Previously I concocted a common denominator and then substituted $x = 1$ and $x = -1$. Let's now do this at source. If we put $x = 1$ in $\frac{2}{(1-x)(1+x)}$ the world (or at least your maths teacher) will explode because it looks as if we're going to have to divide by zero! As you all know, this is the worst sin in mathematics land and condemns the perpetrator to instant mathematical Hades.

So let's hide our sin by covering up the $1 - x$!

Putting $x = 1$ in $\frac{2}{(\!\times\!)(1+x)}$ gives $\frac{2}{(\!\times\!)(1+1)} = 1.$

This turns out to be the coefficient of $\frac{1}{1-x}$ in the partial fraction decomposition of $\frac{2}{1-x^2}$.

Do the same covering up with $x = -1$. We get $\frac{2}{(1-(-1))(\!\times\!)} = 1.$

Again 1 is the coefficient of $\frac{1}{1+x}$ in the decomposition of $\frac{2}{1-x^2}$.

So $\frac{2}{1-x^2} = \frac{1}{1-x} + \frac{1}{1+x}.$

Let's repeat that on $\frac{1}{x(x-1)}$. Now clearly $\frac{1}{x(x-1)} = \frac{a}{x} + \frac{b}{x-1}$ and we can do this the legitimate way that I showed you earlier. But let's live dangerously. Try the cover up method. To find a we let $x = 0$. So we see that $\frac{1}{(\!\times\!)(0-1)} = -1 = a.$

To find b we let $x = 1$. So we get $\frac{1}{1(\!\times\!)} = 1.$ So $b = 1.$

Is it really true that $\frac{1}{x(x-1)} = -\frac{1}{x} + \frac{1}{x-1}$?

Exercises

15. Check that the cover up method did work on $\frac{1}{x(x-1)}$.
16. Repeat Exercises 12 and 13 using the cover up method.

I said that there were two things we had to do with partial fractions. Actually we have done both. We now have found out how to find them and we've also found a use for them. Well you should have done in Exercises 10 and 11. It turns out that a cancelling mechanism comes into play in some examples in the same way that it did when we wanted to find $\sum_{r=1}^{n} r^6$.

To show you what's going on here, I'll do Exercise 11 in full now.

Exercise 11. Show that $\sum_{r=3}^{\infty} \frac{1}{r(r+1)} = \frac{1}{3}$.

The first thing to observe is that $\frac{1}{r(r+1)} = \frac{1}{r} - \frac{1}{r+1}$.

So

$$\sum_{r=3}^{\infty} \frac{1}{r(r+1)} = \sum_{r=3}^{\infty} \left[\frac{1}{r} - \frac{1}{r+1} \right]$$

$$= \left(\frac{1}{3} - \frac{1}{4} \right) + \left(\frac{1}{4} - \frac{1}{5} \right) + \left(\frac{1}{5} - \frac{1}{6} \right) + \cdots .$$

The "$\frac{-1}{4}$" cancels out with the "$\frac{1}{4}$", the "$\frac{-1}{5}$" with the "$\frac{1}{5}$" and so on. For each $\frac{-1}{s}$ term there is a $\frac{1}{s}$ term next to it that cancels it. Hence

$$\sum_{r=3}^{\infty} \frac{1}{r(r+1)} = \frac{1}{3}.$$

Exercises

17. Find the sums of the following.

(i) $\dfrac{1}{6} + \dfrac{1}{12} + \dfrac{1}{20} + \cdots + \dfrac{1}{(r+1)(r+2)} + \cdots + \dfrac{1}{90}$;

(ii) $\dfrac{1}{8} + \dfrac{1}{24} + \dfrac{1}{48} + \cdots + \dfrac{1}{4r(r+1)} + \cdots + \dfrac{1}{840}$;

(iii) $\dfrac{1}{3} + \dfrac{1}{15} + \dfrac{1}{35} + \cdots + \dfrac{1}{(2r+1)(2r-1)} + \cdots + \dfrac{1}{399}$;

(iv) $\dfrac{3}{18} + \dfrac{3}{54} + \dfrac{3}{108} + \cdots + \dfrac{3}{3r(3r+3)} + \cdots + \dfrac{3}{1188}$.

18. Find the following sums.

(i) $\sum_{r=1}^{\infty} \frac{1}{r(r+1)}$;　(ii) $\sum_{r=1}^{\infty} \frac{1}{r(r+2)}$;

(iii) $\sum_{r=1}^{\infty} \frac{1}{r(r+3)}$;　(iv) $\sum_{r=2}^{\infty} \frac{1}{r^2-1}$.

Why are the answers to (ii) and (iv) the same?

19. A sequence of numbers a_n, for $n \geq 1$ is defined as follows: $a_1 = \frac{1}{2}$ and $a_n = \left(\frac{2n-3}{2n} \right) a_{n-1}$. Prove that $\sum_{k=1}^{n} a_k < 1$ for all $n \geq 1$.

1.3. Geometric Progressions

Now you all know that $\frac{1}{3} = 0.3333\ldots$ and that when we get recurring decimals we often use a dot notation $(0.\dot{3})$ to describe them. But did you know, is it in fact true, that every fraction gives a recurring decimal or repeating decimal?

You can chew over that one for a while, while I turn things round the other way. Is every repeating decimal a fraction? Putting these two things together, is it a necessary and sufficient condition for a number to be a fraction, that its decimal form is repeating?

Exercise

20. Which of the following fractions have a recurring decimal representation?
    (i) $\frac{1}{6}$; (ii) $\frac{1}{9}$; (iii) $\frac{1}{11}$; (iv) $\frac{1}{5}$; (v) $\frac{1}{7}$.

Ah! $\frac{1}{10}$ doesn't have a repeating decimal. After all, $\frac{1}{10} = 0.1$. The decimal stops dead. Unless... Unless we write 0.1 as $0.1\dot{0}$!

Some people say that numbers like 0.42 or 0.76543 or 0.1 are **terminating** decimals. However, let's make an agreement between friends, that terminating decimals are in fact repeating decimals with 0 repeating at the end. So $0.42 = 0.42\dot{0}$ and $0.76543 = 0.76543\dot{0}$.

Now then, is a number a fraction if and only if it has a repeating decimal representation?

Exercises

21. Which of the following fractions have a recurring decimal representation?
    (i) $\frac{1}{17}$; (ii) $\frac{1}{25}$; (iii) $\frac{1}{777}$; (iv) $\frac{221}{999}$.
22. What remainder is left when 27 is divided by 83? Keep dividing 27 by 83. What is the reminder after the second division? What is the remainder after the third division? Keep dividing. Are any of the remainders equal? Why? What does this mean for the fraction $\frac{27}{83}$?
23. Prove that if m, $n \in \mathbb{N}$, then $\frac{m}{n}$ can be expressed as a recurring decimal.
24. Does it work the other way? Suppose we are given a recurring decimal out of the blue. Can we find out if it comes from a fraction or not?
    Suppose we had one of those terminating decimals. Is **it** a fraction?

Let's now take a look again at $0.\dot{3}$.

Let $S = 0.\dot{3}$. Then

$$S = 0.3 + 0.03 + 0.003 + \cdots$$

$$= \frac{3}{10} + \frac{3}{100} + \frac{3}{1000} + \cdots$$

Is there some way of getting rid of all those little fractions?

$$10S = 3 + \frac{3}{10} + \frac{3}{100} + \frac{3}{1000} + \cdots .$$

Now look. On the right side of that last equation we've really got S again.

$$10S = 3 + S.$$

So

$$9S = 3 \quad \text{or} \quad S = \frac{1}{3}.$$

We've now seen how to get rid of the endless set of fractions (the repeating decimal part) so that S falls out as a fraction. Does this always work?

Example 1. Show that $0.\dot{7}\dot{1} = 0.717171\ldots$ is a fraction.

Try the same thing again. Let $S = 0.\dot{7}\dot{1}$. Then

$$S = \frac{71}{100} + \frac{71}{10000} + \frac{71}{1000000} + \frac{71}{100000000} + \cdots$$

$$= 71 \times 10^{-2} + 71 \times 10^{-4} + 71 \times 10^{-6} + 71 \times 10^{-8} + \cdots$$

What should we multiply by this time? How about 10? This would give

$$10S = 71 \times 10^{-1} + 71 \times 10^{-3} + 71 \times 10^{-5} + 71 \times 10^{-7} + \cdots$$

Nothing on the right-hand side looks like S. Try 10^2 then. So

$$10^2 S = 71 + 71 \times 10^{-2} + 71 \times 10^{-4} + 71 \times 10^{-6} + \cdots .$$

Ah, that's better.

$$10^2 S = 71 + S.$$

So $99S = 71$, which gives $S = \frac{71}{99}$. That looks like a fraction to me.

Exercises

25. Show that the following recurring decimals are fractions.
    (i) $0.\dot{1}$; (ii) $0.\dot{7}$; (iii) $0.\dot{2}\dot{1}$; (iv) $0.\dot{2}0\dot{1}$.

26. Show that **every** repeating decimal is a fraction, or find a repeating decimal which isn't a fraction.

If we look a little more closely at repeating decimals we see something interesting. Look at $0.\dot{3}$, for instance. Here we have $\frac{3}{10} + \frac{3}{100} + \frac{3}{1000} + \frac{3}{10000} + \cdots$. To get the next term what do we have to do? Multiply by $\frac{1}{10}$. This goes on and on for ever.

The same sort of thing happens with $0.\dot{7}\dot{1}$, though here we multiply by $\frac{1}{100}$ each time. Let's generalise that. Clearly we can look at numbers that change by a factor of 10^{-3} or 10^{-4} and so on. However, we can do things more generally. Why don't we multiply by some number r? So we get, say

$$3 + 3r + 3r^2 + 3r^3 + 3r^4 + \cdots + 3r^{n-1} + \cdots$$

But more generally still, start with a as the first term, not 3. Then we get

$$a + ar + ar^2 + \cdots + ar^{n-1} + \cdots$$

How can we sum this infinite expression? Is it a fraction? Will a sum always exist?

Incidentally, an expression like $\sum_{n=1}^{\infty} ar^{n-1}$, which goes on for ever, is said to be the sum of an **infinite geometric progression**. The a is called the **first term** and the r is the **common ratio** (between consecutive terms).

The sum $\sum_{n=1}^{\infty} ar^{n-1}$ is the same as $a + ar + ar^2 + ar^3 + \cdots$. The sum is considered to be going on for ever. Hence you never actually get to put $n = \infty$ in the summation. The ∞ at the top of the $\sum$ just warns you that there's an awful lot of adding to do.

Exercises

27. Let $S = \sum_{n=1}^{\infty} ar^{n-1}$ where $a = 2$ and $r = \frac{1}{3}$. Using the technique of Example 1, express S as a fraction.

    (The "technique of Example 1" was multiply by 100. What corresponds to 100 in this exercise?)

28. Express the following as fractions.
    (i) $S = \sum_{n=1}^{\infty} \left(\frac{1}{4}\right)^{n-1}$; (ii) $S = \sum_{n=1}^{\infty} \left(\frac{2}{3}\right)^{n-1}$.

29. Express the following as fractions.

(i) $2 + \dfrac{2}{5} + \dfrac{2}{25} + \dfrac{2}{125} + \cdots$; (ii) $3 + \dfrac{3}{7} + \dfrac{3}{49} + \dfrac{3}{343} + \cdots$.

30. Use the technique of Example 1 to find a simple form for S, where
$$S = a + ar + ar^2 + \cdots + ar^{n-1} + \cdots.$$
Does your answer hold for **all** values of r? If not, what happens to other values of r? If so, prove it.

1.4. Extending the Binomial Theorem

We now know that $S = \sum_{n=1}^{\infty} ar^{n-1}$ is equal to $\frac{a}{1-r}$, provided $|r| < 1$. Put $a = 1$ and $r = x$. This gives

$$\frac{1}{1-x} = 1 + x + x^2 + x^3 + x^4 + \cdots$$

So we have a formal expansion for $(1-x)^{-1}$. The Binomial Theorem (see Chapter 2 of *First Step*) tells us how to form $(1-x)^n$ for $n \in \mathbb{N}$, but it doesn't tell us anything about n a negative number. Have we found a generalisation?

Exercises

31. Using the expansion for $(1-x)^{-1}$ as a starting point, find formal expansions for
    (i) $(1-2x)^{-1}$; (ii) $(1+x)^{-1}$; (iii) $(1-x^2)^{-1}$.
32. Check (iii) by directly multiplying together $(1-x)^{-1}$ and $(1+x)^{-1}$. Also check it using partial fractions.

Now, if you remember the Binomial Theorem, you will know that

$$(1+x)^n = \sum_{r=0}^{n} {}^nC_r x^r = \sum_{r=0}^{n} \frac{n!}{r!(n-r)!} x^r.$$

What happens if we put $n = -1$ in this expansion of $(1+x)^n$? Straight away we get problems because $n!$ becomes $(-1)!$. We don't know what $(-1)!$ is.

Let's back off a little. The expression $\frac{n!}{r!(n-r)!}$ cancels a little to give $\frac{n(n-1)\cdots(n-r+1)}{r!}$. Can we put $n = -1$ into this? Yes. That's not so bad. We get $\frac{(-1)(-2)\cdots(-r)}{r!}$. In fact that's $\frac{r!}{r!} = 1$ if r is even, but its $\frac{-(r!)}{r!} = -1$ when r is odd.

So for $r = 0$ we get 1, for $r = 1$ we get -1, for $r = 2$ it's 1, for $r = 3$ it's -1 and so on. Now that's really quite interesting because we've already found (see Exercise 31) that

$$\frac{1}{1+x} = 1 - x + x^2 - x^3 + \cdots .$$

In this expansion the coefficients of x are alternately 1 and -1. Maybe the generalisation that we want for $(1+x)^n$ for n a negative number is

$$(1+x)^n = \sum_{r=0}^{n} \left[\frac{n(n-1)\cdots(n-r+1)}{r!} \right] x^r .$$

That can't be right. How is r ever going to get to n, a negative number? What happens in the expansion for $n = -1$ is that r just keeps getting bigger and bigger. So perhaps:

$$(1+x)^n = \sum_{r=0}^{\infty} \left[\frac{n(n-1)\cdots(n-r+1)}{r!} \right] x^r . \qquad (1)$$

Exercises

33. Expand the expressions of Exercise 31 using equation (1). Do they agree with what you got last time?
34. Expand the following using equation (1).
    (i) $(1+x)^{-2}$; (ii) $(1+x)^{-3}$; (iii) $(1+x)^{-4}$.
35. If you know how to differentiate, check your answers to the last exercise by differentiating $(1+x)^{-1}$.
36. We now know that equation (1) is true and an extension of the Binomial Theorem.
    Find the formal expansion of the following:
    (i) $(1-x^2)^{-2}$; (ii) $\dfrac{x}{1-x}$; (iii) $\dfrac{1}{x(1-x)}$;
    (iv) $(1+x^2)^{-3}$; (v) $\dfrac{1}{2-x}$; (vi) $\dfrac{1}{4-x^2}$;
    (vii) $\dfrac{1}{1-4x^2}$; (viii) $(1-9x^2)^{-4}$; (ix) $\dfrac{1}{(1+x)(1+2x)}$.

1.5. Recurrence Relations

Fibonacci has a lot to answer for. He fancied himself as an applied mathematician when he set out to model, quite unrealistically, rabbit populations in the Pisa area in Italy. What he ended up with was the sequence $1, 1, 2, 3, 5, 8, \ldots$. You've probably met the ***Fibonacci sequence*** before. To find the next term in the sequence you all know that you add the last two terms. So the sequence goes on $13, 21, 34$, etc.

The big question is, what is the n-th term of the Fibonacci sequence? OK, so, the first term is $a_1 = 1$, the second $a_2 = 1$, then $a_3 = 2, a_4 = 3$, $a_5 = 5, a_6 = 8$ and so on. But what is a_n?

At the moment that's too hard to answer directly. We do know though, that for any given value of n we could work it out.

Exercises

37. Find the n-th term for the Fibonacci sequence where $n = 10, 20, 50$, $100, 200$.
38. Suppose we start the Fibonacci sequence with 1, 2 instead of 1, 1. Write down the first eight terms of this Fibonacci-like sequence.
39. Prove that the sequence of the last exercise has no consecutive even numbers. Does it have two consecutive odd numbers?
40. For what n in the Fibonacci sequence, is a_n even?

Now the Fibonacci sequence is such that

$$a_n = a_{n-1} + a_{n-2}. \tag{2}$$

However, this is not enough to define the sequence uniquely. We get different sequences depending on the starting values a_1 and a_2. (We saw this in Exercise 38.)

The Fibonacci sequence though, *is* defined by equation (2) *and* $a_1 = 1$, $a_2 = 1$.

Equation (2) is called a ***recurrence relation***. It's a relation between the terms of the sequence which continually recurs. Given a recurrence relation and the right number of starting values, ***initial values***, a unique sequence of numbers can always be generated.

Exercises

41. Give the first five terms of the sequences produced by the following recurrence relations.
    (i) $a_n = 2a_{n-1}$, $a_1 = 3$;
    (ii) $a_n = 4a_{n-2}$, $a_1 = 1$;
    (iii) $a_n = a_{n-1} + 5$, $a_1 = 3$;
    (iv) $a_n = a_{n-1}^2 - 4$, $a_1 = 1$;
    (v) $a_n = a_{n-1}^2 + a_{n-2}$, $a_1 = 0$, $a_2 = 1$;
    (vi) $a_n = 2a_{n-1} + a_{n-2}$, $a_1 = 1 = a_2$;
    (vii) $a_n = a_{n-1} - a_{n-2}$, $a_1 = 1 = a_2$;
    (viii) $a_n = a_{n-2} - a_{n-1}$, $a_1 = 1 = a_2$.

42. Find recurrence relations for the following sequences.
     (i) $5, 10, 20, 40, 80, \ldots$; (ii) $5, 15, 45, 135, 405, \ldots$;
     (iii) $5, 7, 9, 11, 13, \ldots$; (iv) $5, 7, 12, 19, 31, \ldots$;
     (v) $5, 7, 17, 31, 65, \ldots$; (vi) $5, 7, 24, 62, 172, \ldots$.

What would be nice would be a method of solving recurrence relations so that we could find the value of the n-th term. This would make it easy then to calculate, say, the 594th term of the Fibonacci sequence. It would save us adding term to term via equation (2) until we got as far as 594. It would also mean that next week, when we wanted the 638th term, we wouldn't have to use the recurrence relation and start all over again.

Finding the n-th term of the Fibonacci sequence is not too easy. So let's work up to it. Let's try to find an expression for the general term of the sequence which has $a_1 = 1$ and recurrence relation $a_n = 2a_{n-1}$.

First let us see what the first few terms look like. If $a_1 = 1$, then $a_2 = 2 \times a_1 = 2$. So $a_3 = 2 \times a_2 = 4$. The sequence must be $1, 2, 4, 8, 16, \ldots$. So what is a_n (apart from being $2a_{n-1}$)? It's clearly some power of 2. Is it 2^n or 2^{n+1} or something close to one of these?

If $a_n = 2^n$, then $a_1 = 2$. Clearly $a_n = 2^{n+1}$ won't work either. It's always too big. What if we try something smaller like $a_n = 2^{n-1}$? Then $a_1 = 2^0 = 1$, $a_2 = 2^1 = 2$, $a_3 = 2^2 = 4$ and so on. That looks pretty good.

Exercise

43. Guess the n-th term of the sequences defined by the recurrence relations below.
     (i) $a_n = 2a_{n-1}, a_1 = 2$; (ii) $a_n = 3a_{n-1}, a_1 = 1$;
     (iii) $a_n = 4a_{n-1}, a_1 = 3$; (iv) $a_n = 5a_{n-1}, a_1 = -1$.

But in the sequence defined by $a_n = 2a_{n-1}$ with $a_1 = 1$, how can we be absolutely sure that $a_n = 2^{n-1}$? It certainly fits the pattern alright. But maybe, after starting nicely, it runs off the rails. Remember the number of regions into which lines divide a circle? That looked easy in Chapter 6 of *First Step* till we tried $n = 6$, and then the 2^{n-1} pattern blew up in our faces.

So, if it's true here that $a_n = 2^{n-1}$, how can we *prove* it? Do we have the tools? Well, something like this suggests we try mathematical induction (again see Chapter 6).

Claim. *If $a_n = 2a_{n-1}$ with $a_1 = 1$, then $a_n = 2^{n-1}$.*

Proof. Step 1. Check the case $n = 1$. Now $a_1 = 1$ — we were told this. Further $2^{1-1} = 2^0 = 1$. So $a_n = 2^{n-1}$ for $n = 1$.

Step 2. Assume $a_k = 2^{k-1}$ for some value k.

Step 3. We now have to show that $a_{k+1} = 2^k$. But $a_{k+1} = 2a_k$, by the recurrence relation. By Step 2 we know that we can assume that $a_k = 2^{k-1}$. Hence $a_{k+1} = 2.2^{k-1} = 2^k$ — just what we wanted!

So unlike the regions in the plane, this time things have worked out. We now know for sure (it's no longer a conjecture) that if $a_n = 2a_{n-1}$ with $a_1 = 1$, then $a_n = 2^{n-1}$.

Exercises

44. Prove that your conjectures for Exercise 43 are correct (or guess again).
45. Conjecture the value of the n-th term of each of the following sequences. Prove your conjectures are true using mathematical induction.
    (i) $a_n = 2a_{n-1}, a_1 = 5$;   (ii) $a_n = a_{n-1} + 3, a_1 = 5$;
    (iii) $a_n = a_{n-1} + 5, a_1 = 5$;   (iv) $a_n = a_{n-1} - 2, a_1 = 5$;
    (v) $a_n = 2a_{n-1} + 1, a_1 = 5$;   (vi) $a_n = 3a_{n-1} - 2, a_1 = 4$.

So far we have seen how to use recurrence relations to produce a sequence of numbers. Now sometimes a sequence of numbers is the answer to a problem. For instance, how many subsets are there of a set of size n?

One way of answering that question is to find a_1 (the number of subsets of a set with one element) and a recurrence relation for a_n (the number of subsets of a set with n elements). Then, if we're lucky, we can use the techniques of Exercises 44 and 45 to solve the recurrence relation for a_n. This will then solve our subset problem.

OK, so what is a_1? Let $S_1 = \{1\}$. How many subsets does it have? Well $\emptyset \subseteq S_1$ and $\{1\} \subseteq S_1$ That's all, so $a_1 = 2$.

Before we can get at a_n, it'll probably be useful to write out a few specific terms. Suppose we look at a_2, a_3 and a_4. With many counting problems it's a good idea to count the small values first. It gives you a feel for what it is you are actually counting and it gives you some evidence to build on for the big n-th term.

So let $S_2 = \{1, 2\}$. Now the subsets here are $\emptyset, \{1\}, \{2\}, \{1, 2\} = S_2$. There are four of them, so $a_2 = 4$.

Let $S_3 = \{1, 2, 3\}$. Here the subsets are $\emptyset, \{1\}, \{2\}, \{3\}, \{1, 2\}, \{1, 3\}, \{2, 3\}, \{1, 2, 3\} = S_3$. Hence $a_3 = 8$.

Now if we're trying to find a recurrence relation for a_n, then somehow we've got to build up a_n from a_{n-1} or smaller terms. Can we find a_4 by building up from the subsets of S_3?

Let $S_4 = (1, 2, 3, 4)$. Clearly any subset of S_3 is a subset of S_4. (This means that $a_4 \geq a_3 = 8$, at least.) What subsets of S_4 are **not** already subsets of S_3? Surely they are the ones with a 4 in them. These sets are $\{4\}, \{1, 4\}, \{2, 4\}, \{3, 4\}, \{1, 2, 4\}, \{1, 3, 4\}, \{1, 2, 3, 4\} = S_4$. There are eight of these subsets too. Is that a coincidence? Is there a link between the subsets of S_3 and the subsets of S_4 which contain 4? Let's list them to see:

Subsets of S_3 are $\emptyset, \{1\}, \{2\}, \{3\}, \{1, 2\}, \{1, 3\}, \{2, 3\}, \{1, 2, 3\}$

Subsets of S_4 containing 4 are $\{4\}, \{1, 4\}, \{2, 4\}, \{3, 4\}, \{1, 2, 4\}, \{1, 3, 4\}, \{2, 3, 4\}, \{1, 2, 3, 4\}$.

You should begin to see some pattern here. To get the subsets with 4 in them all you have to do is to add 4 to the subsets without a 4 in them.

So if Y is a subset of S_4, then either $4 \in Y$ or $4 \notin Y$. If $4 \notin Y$ then $Y \subseteq S_3$. On the other hand, if $4 \in Y$, then $Y \backslash (4) \subseteq S_3$. Since adding 4 to different subsets of S_3 gives different subsets of S_4 and removing 4 from a subset of S_4 gives a subset of S_3, there are as many subsets of S_4 which contain 4 as there are which don't. There are a_3 which don't. So $a_4 = 2a_3 = 16$.

But this argument holds for $S_n = \{1, 2, 3, \ldots, n\}$. Let $Y \subseteq S_n$. Then either $n \notin Y$ or $n \in Y$. If $n \notin Y$, then $Y \subseteq S_{n-1}$. If $n \in Y$, then $Y \backslash \{n\} \subseteq S_{n-1}$. Since each $Y \subseteq S_{n-1}$ gives a different subset $Y \cup \{n\}$ of S_n which contains n and vice-versa, then the number of subsets of S_n which contain n, equals the number of subsets of S_{n-1}. Hence $a_n = 2a_{n-1}$.

Having got our recurrence relation we can solve for a_n.

Exercises

46. How many subsets does a set of n elements have?
47. Let a_n be the number of n-digit numbers that can be formed using only the digits $1, 2, 3, 4$. Find a_1, a_2 and a recurrence relation for a_n. Use this recurrence relation to find a_n.
48. Find a_n, the number of words (mostly nonsense words) that can be formed by the letters A, B, C, D, E.
49. Let b_n be the number of n-digit numbers made from the numbers $1, 2, 3, 4$, which have an odd number of l's. Find b_1, b_2 and a recurrence relation for b_n.

    Solve the recurrence relation.

    Hence find a formula for c_n, the number of n-digit numbers made from $1, 2, 3, 4$ which have an even number of l's.

So we seem to be able to manage to solve recurrence relations like $a_n = Aa_{n-1}$. There are bigger problems, though, with things like $a_n = Aa_{n-1} + Ba_{n-2}$. To solve these we need a theorem.

Theorem 1. *Let $a_n = Aa_{n-1} + Ba_{n-2}$ where A and B are known constants.*
(a) *If $x^2 = Ax + B$ has distinct roots α and β then $a_n = K\alpha^n + L\beta^n$, where K and L are constants which can be found in terms of a_1 and a_2.*
(b) *If $x^2 = Ax + B$ has the repeated root α, then $a_n = (K + nL)\alpha^n$, where K and L are constants which can be found in terms of a_1 and a_2.*

At first sight, the theorem doesn't look very pretty. So let's have a look at a couple of examples to see how it works: it's really pretty harmless.

Example 2. Find the general term of the sequence defined by the recurrence relation

$$a_n = 3a_{n-1} - 2a_{n-2} \quad \text{where } a_1 = 1 \text{ and } a_2 = 2.$$

It's always worth getting a feel for these things by working out a few terms.

Since $a_1 = 1$ and $a_2 = 2$, it follows that $a_3 = 3 \times 2 - 2 \times 1 = 4$, $a_4 = 3 \times 4 - 2 \times 2 = 8$, $a_5 = 3 \times 8 - 2 \times 4 = 16$, and so on. Surely $a_n = 2^{n-1}$!
    The theorem says that since $a_n = 3a_{n-1} - 2a_{n-2}$, i.e. $A = 3$ and $B = -2$, we have to look at the quadratic $x^2 = 3x - 2$. Now this gives $x^2 - 3x + 2 = 0$. So $(x - 1)(x - 2) = 0$. The roots of the quadratic are therefore $\alpha = 1$ and $\beta = 2$.
    The theorem then tells us, that since $\alpha \neq \beta$, it follows that $a_n = K1^n + L2^n$ and that we can find K and L from a_1 and a_2.
    Starting with $a_n = K1^n + L2^n$, first put $n = 1$ and then put $n = 2$. This gives

$$1 = a_1 = K + 2L,$$
$$2 = a_2 = K + 4L.$$

Solving these two equations for K and L we get $K = 0$ and $L = 1/2$. Hence

$$a_n = 0 \times 1^n + 1/2 \times 2^n$$

i.e.

$$a_n = 2^{n-1}.$$

Just what we guessed!

Example 3. Find the general term of the sequence defined by the recurrence relation

$$a_n = 2a_{n-1} - a_{n-2} \quad \text{where } a_1 = 1 \text{ and } a_2 = 2.$$

Again, try to find a few terms. This won't always help us with the general term but it will give us some idea of what sequence it is we're dealing with. And it will provide a useful check.

So $a_3 = 2 \times 2 - 1 = 3, a_4 = 2 \times 3 - 2 = 4, a_5 = 2 \times 4 - 3 = 5$. It looks like $a_n = n$. But can that be right?

What does the theorem say? In the present case $A = 2$ and $B = -1$. So we have to solve the quadratic $x^2 = 2x - 1$. Rearranging we get $x^2 - 2x + 1 = 0$ or $(x-1)^2 = 0$. The only root of this equation is 1. Thus we use the second part of the theorem with $\alpha = 1$ to find that

$$a_n = (K + nL)1^n.$$

As before we find K and L by using a_1 and a_2.

$$1 = a_1 = K + L,$$

$$2 = a_2 = K + 2L.$$

Solving gives $K = 0$ and $L = 1$. So $a_n = n$. How about that?

Exercises

50. Use the theorem to find the general term of the sequences described below.
    (i) $a_n = 4a_{n-1} - 3a_{n-2}, a_1 = 1, a_2 = 1$;
    (ii) $a_n = 4a_{n-1} - 3a_{n-2}, a_1 = 3, a_2 = 9$;
    (iii) $a_n = 4a_{n-1} - 3a_{n-2}, a_1 = 1, a_2 = 9$;
    (iv) $a_n = 5a_{n-1} - 6a_{n-2}, a_1 = 0, a_2 = 1$;
    (v) $a_n = 4a_{n-1} - 4a_{n-2}, a_1 = 0, a_2 = 1$.

51. Use the theorem to find the general term of the Fibonacci sequence.

52. Let a_n be the number of n-digit numbers that can be made using only 1 and 2, if no consecutive 2's are allowed. Show that $a_1 = 2, a_2 = 3$ and $a_3 = 5$. Find a recurrence relation for a_n.

53. Let a_n be the number of ways of hanging red and white shirts on a line so that no two red shirts are next to each other. Find a recurrence relation for a_n.

54. Let a_n be the number of n-digit numbers that can be made using only $1, 2, 3$, if no consecutive 2's and no consecutive 3's are allowed. Find a formula for a_n.

55. How many words of length n can be made using only A, B, C if no two consecutive A's are allowed?

56. Prove the theorem using mathematical induction. (Warning: this is quite difficult — if you can manage this one on your own you are doing very well indeed.)

57. An integer sequence is defined by $a_n = 2a_{n-1} + a_{n-2}$ for $n > 2$ and $a_1 = 1$, $a_2 = 2$. Prove that 2^k divides a_n if and only if 2^k divides n.

1.6. Generating Functions

We saw in the last section that, given the recurrence relation

$$a_n = Aa_{n-1} + Ba_{n-2}$$

we need to use a theorem to find an expression for a_n. In fact we can also tackle recurrence relations through formal power series.

For instance,

$$F(x) = x + x^2 + 2x^3 + 3x^4 + 5x^5 + 8x^6 + 13x^7 + \cdots$$

is a formal power series whose coefficients are the numbers in the Fibonacci sequence. If we can obtain $F(x)$ from the Fibonacci recurrence relation, it may save us using the theorem of the last section. We may also find that we can apply the power series method to recurrence relations that the theorem doesn't cover. Let's see how far we can go. Let

$$f(x) = \sum_{n=1}^{\infty} a_n x^n = a_1 x + a_2 x^2 + a_3 x^3 + \cdots + a_n x^n + \cdots$$

be a formal power series. Then $f(x)$ is said to be the **generating function** for the sequence $a_1, a_2, a_3, \ldots, a_n, \ldots$

The sequence $1, 1, 1, \ldots$ is not the most exciting you've ever seen but its generating function is $g(x) = x + x^2 + x^3 + x^4 + \ldots + x^n$. From Section 1.4 we know that

$$g(x) = \frac{x}{1-x}.$$

So $x(1-x)^{-1}$ is the generating function of $1, 1, 1, \ldots$.

Exercise

58. Find generating functions for the following sequences. Write the generating functions as simply as possible using the results of Section 1.4.

(i) $1, 2, 3, 4, \ldots$; (ii) $1, 0, 1, 0, 1, \ldots$; (iii) $1, -1, 1, -1, 1, \ldots$;
(iv) $1, 3, 6, 10, 15, 21, \ldots$; (v) $2, 4, 8, 16, \ldots$; (vi) $2, -1, 1/2, -1/4, \ldots$.

Now let's see how to use a generating function to solve a recurrence relation.

Example 4. Find the generating function for the sequence with $a_1 = 1$ and $a_n = 2a_{n-1}$.

Let $f(x) = \sum_{n=1}^{\infty} a_n x^n$ be the required generating function. Then

$$f(x) = a_1 x + a_2 x^2 + a_3 x^3 + \cdots + a_n x^n + \cdots$$
$$= a_1 x + 2x(a_1 x + a_2 x^2 + a_3 x^3 + \cdots + a_{n-1} x^{n-1} + \cdots)$$
$$= a_1 x + 2x f(x)$$
$$= x + 2x f(x).$$

Solving this equation for $f(x)$ gives

$$f(x) = \frac{x}{1 - 2x}.$$

But we know, by the extended Binomial Theorem, that

$$(1 - 2x)^{-1} = \sum_{m=0}^{\infty} 2^m x^m,$$

Hence $f(x) = x \sum_{m=0}^{\infty} 2^m x^m = \sum_{m=0}^{\infty} 2^m x^{m+1} = \sum_{n=0}^{\infty} 2^{n-1} x^n$.
Since $f(x) = \sum_{n=1}^{\infty} a_n x^n$, we have discovered that $a_n = 2^{n-1}$. This is consistent with the answer we obtained in Section 1.5.

Exercise

59. Find generating functions for the following sequences and hence find the general term.
   (i) $a_n = 3a_{n-1}, a_1 = 1$; (ii) $a_n = 2a_{n-1}, a_1 = 1$;
   (iii) $a_n = 4a_{n-1}, a_1 = 1$; (iv) $a_n = -3a_{n-1}, a_1 = 1$;
   (v) $a_n = a_{n-1} + 1, a_1 = 1$; (vi) $a_n = a_{n-1} + 2, a_1 = 2$.

Example 5. Find the generating function of the Fibonacci sequence. Use the generating function to find an expression for a_n in terms of n.

Let

$$F(x) = \sum_{n=1}^{\infty} a_n x^n$$

$$= a_1 x + a_2 x^2 + a_3 x^3 + a_4 x^4 + \cdots + a_n x^n + \cdots$$

$$= x + x^2 + (a_2 + a_1)x^3 + (a_3 + a_2)x^4 + \cdots + (a_{n-1} + a_{n-2})x^n + \cdots$$

$$= x + x^2 + (a_1 x^3 + a_2 x^4 + a_3 x^5 + \cdots + a_{n-2} x^n + \cdots)$$

$$\quad + (a_2 x^3 + a_3 x^4 + a_4 x^5 + \cdots + a_{n-1} x^n + \cdots)$$

$$= x + x^2 + x^2(a_1 x + a_2 x^2 + a_2 x^3 + \cdots + a_{n-2} x^{n-2} + \cdots)$$

$$\quad + x(a_2 x^2 + a_3 x^3 + a_4 x^4 + \cdots + a_{n-1} x^{n-1} + \cdots)$$

$$= x + x^2 + x^2 F(x) + x\{F(x) - a_1 x\}$$

$$= x + x F(x) + x^2 F(x).$$

Solving for $F(x)$ gives

$$F(x) = \frac{x}{1 - x - x^2}.$$

Hmm! Now that's all well and good, but how are we going to find the general term of that power series? Actually you should know the answer to this question.

Exercises

60. Find generating functions for each of the following sequences. Where possible find a_n.
    (i) $a_n = 5a_{n-1} - 6a_{n-2}$, $a_1 = 1$, $a_2 = 5$;
    (ii) $a_n = 2a_{n-1} + 3a_{n-2}$, $a_1 = 4$, $a_2 = 5$;
    (iii) $a_n = 4a_{n-1} + 5a_{n-2}$, $a_1 = 2$, $a_2 = 2$;
    (iv) $a_n = a_{n-1} + a_{n-2}$, $a_1 = 0$, $a_2 = 1$.

61. Show that if $a_n = Aa_{n-1} + Ba_{n-2}$, then
    $$f(x) = \frac{a_1 x + (a_2 x - Aa_1)x^2}{1 - (Ax + Bx^2)}.$$
    Check you answers to Exercise 60 using this general form.

Finding the general term for $F(x)$, will involve some messy algebra. I'll avoid it here and leave it for you to do later. To illustrate the method though have a look at this.

Example 6. Find the general term of the power series

$$f(x) = \frac{x}{1 - 5x - 6x^2}.$$

The trick is: partial fractions! You see, $1 - 5x + 6x^2 = (1 - 3x)(1 - 2x)$. So

$$f(x) = \frac{1}{(1 - 3x)(1 - 2x)} = \frac{1}{1 - 3x} - \frac{1}{1 - 2x}.$$

(I used the, shsh!, cover up method.)

By the extended Binomial Theorem we learn that

$$\frac{1}{1 - 3x} = \sum_{n=0}^{\infty} (3x)^n \quad \text{and} \quad \frac{1}{1 - 2x} = \sum_{n=0}^{\infty} (2x)^n.$$

Hence $f(x) = \sum_{n=0}^{\infty} (3^n x^n - 2^n x^n) = \sum_{n=0}^{\infty} (3^n - 2^n) x^n$.
So that means that the coefficient of x^n is, in fact, $3^n - 2^n$.

In general then, we try to use the partial fraction decomposition of $f(x)$. When we've done that, we can expand using the extended Binomial Theorem. If we then gather the x^n terms together, we've won ourselves a_n.

Exercises

62. Use the above method to find an expression for a_n for each of the recurrence relations in Exercise 60.

63. Use generating functions to find the general terms of the following sequences.
    (i) $a_n = 5a_{n-1} - 6a_{n-2}$, $a_1 = 0$, $a_2 = 1$;
    (ii) $a_n = 2a_{n-1} + 3a_{n-2}$, $a_1 = 1$, $a_2 = 2$;
    (iii) $a_n = 4a_{n-1} + 5a_{n-2}$, $a_1 = 1$, $a_2 = 4$;
    (iv) $a_n = 4a_{n-1} + 5a_{n-2}$, $a_1 = 1$, $a_2 = 3$;
    (v) $a_n = 2a_{n-1} + 3$, $a_1 = 1$;
    (vi) $a_n = 3a_{n-1} + 2$, $a_1 = 2$.

64. Use the generating function $F(x) = \frac{x}{1-x-x^2}$ to find the general term of the Fibonacci sequence.

65. Colour the squares of a $1 \times n$ chessboard either red or white or blue. How many ways are there of doing this if no two white squares are adjacent?

66. I have given you two methods to solve recurrence relations. Are they really the same method? Is one quicker than the other? Is one more useful than the other?

67. There are many interesting features of the Fibonacci numbers. For instance:
    (a) show that $a_{i+j} = a_i a_{j-1} + a_{i+1} a_j$;
    (b) for arbitrary k and n show that a_{kn} is divisible by a_n;
    (c) for arbitrary n, show that a_n and a_{n+1} are relatively prime.
68. Find the greatest common divisor of a_{1990} and a_{2000}, where a_{1990} and a_{2000} are Fibonacci numbers.

1.7. Of Rabbits and Postmen

Fibonacci was trying to model rabbit populations when he ended up with $1, 1, 2, 3, 5, 8, 13, \ldots$. How on earth did he manage that? The story goes this way.

Take one pair of rabbits. Now I'm going to assume that this pair contains one male and one female. Everybody knows that rabbits multiply but they don't multiply immediately. It's certainly not on the minds of newborn pairs who we will call A-pairs. After a month, a newborn A-pair matures into a B-pair. At this stage I have a better idea of what's on their minds because at the end of the next month they mature into a C-pair and have a happy event, the arrival of an A-pair.

From then on the cycle continues. The C-pair produce an A-pair, on cue, every month. After each month an A-pair becomes a B-pair while a B-pair becomes a C-pair with the obvious result — another A-pair.

To give you some idea of what's going on. I've tried to illustrate things with a diagram.

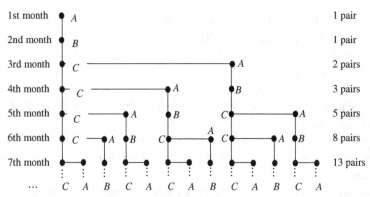

You can see from the column on the right-hand side of the diagram that the number of pairs after month n is beginning to look like a_n, where a_n is the n-th term of the Fibonacci sequence.

But you should be asking yourselves, or else I haven't been bringing you up right, will this continue to be the Fibonacci sequence? Certainly $a_1 = 1 = a_2$, if we let a_n be the number of pairs at the n-th month. However, is it true that $a_n = a_{n-1} + a_{n-2}$? Let's try to find a rabbit recurrence relation.

Clearly $a_n = a_n(A) + a_n(B) + a_n(C)$, where $a_n(A), a_n(B), a_n(C)$ are, respectively, the number of A-, B-, C-pairs in the n-th month.

Where do A-pairs come from? I think you'd better ask your mother or father.

Try that another way. How many A-pairs are there in any month? Well A-pairs are only A-pairs once. Further, there's always one A-pair for every C-pair. Hence

$$a_n(A) = a_n(C).$$

How do we get B-pairs? I can answer that one. We can only get B-pairs this month from A-pairs last month. Any pair of rabbits is a B-pair precisely once — the month after they're born. So

$$a_n(B) = a_{n-1}(A).$$

For what it's worth then

$$a_n(B) = a_{n-1}(A) = a_{n-1}(C).$$

Now for the hard question. Where do C-pairs come from? Surely this happens in two ways. Either a B-pair grows up to be a C-pair or a C-pair stays a C-pair. So this month's C's are last month's C's plus last month's B's. This means that

$$a_n(C) = a_{n-1}(B) + a_{n-1}(C).$$

Now let's put all that together.

$$
\begin{aligned}
a_n &= a_n(A) + a_n(B) + a_n(C) \\
&= \{a_n(C) + a_{n-1}(A)\} + [a_{n-1}(B) + a_{n-1}(C)] \\
&= \{[a_{n-1}(B) + a_{n-1}(C)] + a_{n-1}(A)\} + [a_{n-1}(B) + a_{n-1}(C)] \\
&= a_{n-1} + [a_{n-2}(A) + a_{n-1}(C)] \\
&= a_{n-1} + [a_{n-2}(A) + a_{n-2}(B) + a_{n-2}(C)] \\
&= a_{n-1} + a_{n-2}.
\end{aligned}
$$

Fibonacci rules! So rabbits which behave in the way we've described (from A to B to C), reproduce Fibonaccially.

Exercises

69. When did Fibonacci postulate this model for rabbit growth? Are you impressed with it as a potential model for rabbit populations? Why is your house not completely filled with rabbits? Propose and test a better model.

70. Where did Fibonacci live? Where did he go to school? Did his wife raise rabbits? What did Fibonacci do when he wasn't thinking about our furry friends?

Which, quite naturally, brings us on to crazy postmen. (Well it might be quite naturally for some people.)

There's a new postman in the neighbourhood. Unfortunately he has a bad habit of delivering the letters to the wrong houses. In fact in a street with n houses, each house has been mailed a letter but he put every single letter in the wrong box!

In how many ways can he do this?

Exercises

71. Let p_n be the number of ways the postie delivers n letters to n houses so that no letter gets to the right house. (Remember only one letter was addressed to each house.)
    (a) Find p_1, p_2, p_3, p_4.
    (b) Guess what p_6 is.
    (c) Now actually find p_5, p_6. Is there any pattern here?

72. What we're clearly looking for is a recurrence relation for p_n. If we try to get something of the form $p_n = Ap_{n-1} + Bp_{n-2}$ what might A and B look like?
    Test out your conjectures on various values of n.
    Prove any conjecture that seems to stand up to these tests.

Let's see what we can do for p_5. If the houses are numbered $1, 2, 3, 4, 5$ we can assume the letters are addressed $1, 2, 3, 4, 5$.

The first way things can go wrong, is if letter 1 goes to house 2 and letter 2 goes to house 1. In other words, letters 1 and 2 get their destinations interchanged. I've shown this in the diagram.

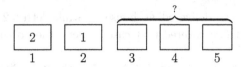

In how many ways can the postman get the letters to houses $3, 4, 5$ the wrong way? It's just 2 isn't it? But more importantly, what does that 2 represent? Isn't it just the number of ways of delivering three letters wrongly to three houses. So we should perhaps really think of 2 as p_3. After all we're trying to get a recurrence relation for p_n. By starting with five houses, we hope to get p_5 in terms of smaller p_n. That way we might see how to get a general recurrence relation.

So we've found p_3 wrong deliveries when 1 and 2 are swapped. How many possible swaps are there with 1? Well, there's 1 and 2, 1 and 3, 1 and 4, 1 and 5. That's four altogether.

If the postman goes wrong by swapping 1 with some other letter, then he can go wrong in $4p_3$ ways.

Exercise

73. The postie is faced with a street with six houses.
    (a) How many completely wrong deliveries can he make if he swaps letter 1 with letter 5?
    Repeat this with *n* houses.
    (b) How many completely wrong deliveries can he make, if he inadvertently swaps letter 1 with another letter in the six house street?
    Generalise.

What then, if letter 1 isn't swapped with any other letter? Suppose letter 1 gets delivered to house 2 but letter 2 doesn't get delivered to house 1. In how many ways can the postman go wrong now?

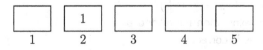

Certainly 3 can't go to 3, 4 can't go to 4 and 5 can't go to 5. What's more 2 can't go to 1. Hmm. What if we readdressed letter 2 as letter 1 for a minute. (That's safe because letter 1 has actually been delivered to house 2.) We now see that the new 1 can't go to 1, 3 can't go to 3, 4 can't got to 4 and 5 can't go to 5. None of the letters $1, 3, 4, 5$ can go to the right house. Surely this can be done in p_4 ways?

So if letter 1 gets delivered at house 2, while letter 2 ***doesn't*** get delivered to house 1 (we covered that case earlier), the postman can work his evil in p_4 ways.

But letter 1 can go to house $2, 3, 4,$ or 5 without having the swap of letters $2, 3, 4$ or 5 going to house 1. So there are four possible destinations for letter 1, if there's no swap. So the no-swap situation produces $4p_4$ more wrong deliveries.

Exercise

74. The postie is faced with the challenge of getting it all wrong with six houses.
    (a) How many completely wrong deliveries in the six house street can he make if he delivers letter 1 to house 4 but letter 4 is not delivered to house 1?
        Repeat this with *n* houses.
    (b) How many completely wrong deliveries in the six house street can he make if he does not swap letter 1 with any other letter.
        Generalise.

But letter 1 must go to some house other than 1. Say it went to house m. Then either letter m goes to house 1 or it doesn't. In the five house street this means that

$$p_5 = 4p_4 + 4p_3. \tag{3}$$

Exercises

75. Conjecture a generalisation to equation (3).
    Prove you generalisation.
76. Write a computer program for p_n for $n \leq 10$.
    Conjecture an expression for p_n in terms of *n*. Can you prove your conjecture?
77. Is $p_3 = 3!(1 - \frac{1}{1!} + \frac{1}{2!} - \frac{1}{3!})$?
    Is $p_4 = 4!(1 - \frac{1}{1!} + \frac{1}{2!} - \frac{1}{3!} + \frac{1}{4!})$?
    Guess a similar expression for p_5 and p_6.
    Prove your guess is correct.
    Generalise.

The number p_n is called the number of ***derangements*** of *n* objects. In terms of permutations it is the number of permutations of *n* things where no object remains in its original position.

The recurrence relation for derangements is, as we have seen above,

$$p_n = (n-1)p_{n-1} + (n-1)p_{n-2}.$$

This recurrence relation does not fit the theorem and method of Section 1.5. This is because $n-1$ is not a constant. Hence it cannot be equated with the A or the B of the theorem. Can p_n be determined by the generating function method though?

1.8. Solutions

1. Gauss: 30 April 1777 to 23 February 1855. The other answers you can find by reading a road map and using the Mactutor web site http://www-history.mcs.st-and.ac.uk/.

2. (i) $a = 1, d = 4$. Since $S_n = \frac{n}{2}[2a + (n-1)d]$, the sum here is 1035; (ii) 2576; (iii) 1840; (iv) 264.5; (v) -414; (vi) -414.

3. (i) Here you need to find the general term in order to find the number of terms altogether. As the terms here go up by 4 each time you are looking at something like $a_n = 4n + c$ for the general term. Trial and error shows that $c = -3$, so $a_n = 4n - 3$. Since the last term is 45, $4n - 3 = 45$, so $n = 12$. Applying the formula gives a sum of 276; (ii) here $a_n = 10n - 8$ and so to get 102 we have $n = 11$. The sum is then 572; (iii) 418; (iv) 1250; (v) -266; (vi) 96.

4. $S_n = \left(\frac{a+\ell}{2}\right) \times n$ for an arithmetic progression.

5. (i) 3; (ii) 11; (iii) 1.

6. (i) $a = -4 - 15n, b = 7 + 26n$; (ii) $a = -1 - 3n, b = 4 + 11n$; (iii) $a = 31 + 72n, b = -65 - 151n$. (In each case, n is any integer.)

7. (i) n; (ii) $\frac{1}{2}n(n+1)$; (iii) $\frac{1}{4}n^2(n+1)^2$; (iv) $\frac{1}{30}n(n+1)(2n+1)(3n^2+3n-1)$. (Now try $\sum r^5$, $\sum r^6$ and $\sum r^7$.) (Sorry for leaving out the algebra needed here.)

8. $\frac{-3}{7} + \frac{1}{2}$; $\frac{4}{7} - \frac{1}{2}$; $\frac{11}{7} - \frac{3}{2}$; and infinitely many others. (Can you find any general form(s)?)

9. (i) $\frac{1}{2} - \frac{2}{5}$; $\frac{3}{5} - \frac{1}{2}$; (ii) $\frac{5}{9} - \frac{1}{2}$; $\frac{3}{2} - \frac{13}{9}$; (iii) $\frac{2}{5} - \frac{1}{3}$; $\frac{2}{3} - \frac{3}{5}$;

   (iv) $\frac{11}{31} - \frac{1}{3}$; $\frac{2}{3} - \frac{20}{31}$; (v) $\frac{3}{7} - \frac{3}{8}$; $\frac{5}{8} - \frac{4}{7}$; (vi) $\frac{2}{5} - \frac{2}{7}$; $\frac{5}{7} - \frac{3}{5}$.

10. $\frac{1}{12} = \frac{1}{3} - \frac{1}{4}$; $\frac{1}{20} = \frac{1}{4} - \frac{1}{5}$; $\frac{1}{30} = \frac{1}{5} - \frac{1}{6}$; $\cdots$; $\frac{1}{9900} = \frac{1}{99} - \frac{1}{100}$.

    Hence $\frac{1}{12} + \frac{1}{20} + \frac{1}{30} + \frac{1}{42} + \cdots + \frac{1}{9900}$

$$= \left(\frac{1}{3} - \frac{1}{4}\right) + \left(\frac{1}{4} - \frac{1}{5}\right) + \left(\frac{1}{5} - \frac{1}{6}\right) + \left(\frac{1}{6} - \frac{1}{7}\right) + \cdots + \left(\frac{1}{99} - \frac{1}{100}\right)$$

$$= \frac{1}{3} - \frac{1}{100} = \frac{97}{300}.$$

11. Now from the last exercise, $\dfrac{1}{r(r+1)} = \dfrac{1}{r} - \dfrac{1}{r+1}$.

So $\sum_{r=3}^{\infty} \dfrac{1}{r(r+1)} = \left(\dfrac{1}{3} - \dfrac{1}{4}\right) + \left(\dfrac{1}{4} - \dfrac{1}{5}\right) + \cdots$

12. (i) $\dfrac{1}{2-x} + \dfrac{1}{2+x}$;   (ii) $\dfrac{3/2}{3-x} + \dfrac{3/2}{3+x}$;   (iii) $\dfrac{2}{4-x} + \dfrac{2}{4+x}$;

    (iv) $\dfrac{1/3}{x-1} - \dfrac{1/3}{x+2}$;   (v) $\dfrac{1}{x+1} - \dfrac{1}{x+2}$;   (vi) $\dfrac{1}{x-1} - \dfrac{1}{x}$.

13. We want to find a, b, c such that $\frac{a}{x} + \frac{b}{x-1} + \frac{c}{x+1}$. So, again put things over a common denominator to get $a(x-1)(x+1) + bx(x+1) + cx(x-1) = 1$. Substituting $x = 0, 1, -1$ should give $a = -1$, $b = \frac{1}{2}, c = \frac{1}{2}$. (Of course you could try the cover up rule!)

14. Perhaps surprisingly, the answers we've got so far look unique. How can we get anything but $a = -1, b = \frac{1}{2}, c = \frac{1}{2}$ using the method we did in the last exercise? Unique it is. The key here is that for partial fractions with linear factors in the denominator, we don't allow anything but constants in the numerator.

15. This should be no problem.

16. You should still get the same answers.

17. (i) $\dfrac{1}{6} + \dfrac{1}{12} + \cdots + \dfrac{1}{90} = \left(\dfrac{1}{2} - \dfrac{1}{3}\right) + \left(\dfrac{1}{3} - \dfrac{1}{4}\right) + \cdots + \left(\dfrac{1}{9} - \dfrac{1}{10}\right)$

$$= \frac{1}{2} - \frac{1}{10} = \frac{2}{5}; \text{ (ii) } \frac{1}{2}\left(\frac{1}{2} - \frac{1}{4}\right) + \frac{1}{2}\left(\frac{1}{4} - \frac{1}{6}\right) + \cdots + \frac{1}{2}\left(\frac{1}{28} - \frac{1}{30}\right) = \frac{7}{30};$$

(iii) $\dfrac{1}{2}\left(1 - \dfrac{1}{21}\right) = \dfrac{10}{21}$;   (iv) $\left(\dfrac{1}{3} - \dfrac{1}{6}\right) + \cdots + \left(\dfrac{1}{33} - \dfrac{1}{36}\right) = \dfrac{11}{36}$.

18. (i) $\dfrac{1}{r(r+1)} = \dfrac{1}{r} - \dfrac{1}{r+1}$, so

$$\sum_{r=1}^{\infty} \frac{1}{r(r+1)} = \left(\frac{1}{1} - \frac{1}{2}\right) + \left(\frac{1}{2} - \frac{1}{3}\right) + \cdots = 1;$$

(ii) $\dfrac{1}{r(r+2)} = \dfrac{\frac{1}{2}}{r} - \dfrac{\frac{1}{2}}{r+2}$ so $\sum_{r=1}^{\infty} \dfrac{1}{r(r+2)} = \dfrac{1}{2}$;   (iii) $\dfrac{1}{3}$;

(iv) $\dfrac{1}{2}$ (as implied by the question).

    Parts (ii) and (iv) are equal because they are the same sum. Substitute $s = r - 1$ in (iv) and see what happens.

19. In the spirit of this section we want to rearrange things so that there is a cancelling. Show that $a_k = 2ka_k - 2(k+1)a_{k+1}$. I'll give you the rest of the solution later.

    [This problem was submitted by Iceland in the 29th IMO in Canberra. It was not used for that Olympiad.]

20. (i) $0.1\dot{6}$; (ii) $0.\dot{1}$; (iii) $0.0\dot{9}$; (iv) 0.2; (v) $0.\dot{1}4285\dot{7}$.

21. (i) Yes; (ii) Yes; (iii) Yes; (iv) Yes.

22. $\frac{27}{83} = 0.2 + \text{rem}.21 = 0.32 + \text{rem}.44 = 0.325 + \text{rem}.25$

    $= 0.3253 + \text{rem}.1 = 0.32530 + \text{rem}.10 = 0.32531 + \text{rem}.17$

    $= 0.325312048 + \text{rem}.16 = 0.3253120481 + \text{rem}.77$

    $= 0.32531204819 + \text{rem}.23 = 0.325312048192 + \text{rem}.64$

    $= 0.3253120481927 + \text{rem}.59 = \cdots$

    Wait a minute. This is all boring and tedious and totally unnecessary. Look at those remainders. Every one of them is naturally under 83. Eventually a remainder will be zero OR a remainder must come up that we've seen before. That'll have to happen in less than or equal to 83 divisions. When it happens the whole sequence will go round again till the same remainder occurs. Then off we go again and again. So $\frac{27}{83}$ must be a repeating decimal.

23. The argument is exactly the same as that of Exercise 22.

24. Yes. Yes. Yes. (But why?)

25. (i) If $S = 0.\dot{1}$, then $10S = 1 + S$. Hence $S = \frac{1}{9}$; (ii) If $S = 0.\dot{7}$, then

    $10S = 7 + S$. Hence $S = \frac{7}{9}$; (iii) $S = \frac{7}{33}$; (iv) $S = \frac{67}{333}$.

26. Surely every repeating decimal is a fraction. But how to prove it? It's fairly obvious how to do each **particular** example. But how to do it in general?

27. $S = 2 + \frac{2}{3} + \frac{2}{9} + \cdots$, then $3S = 6 + 2 + \frac{2}{3} + \frac{2}{9} + \cdots = 6 + S$. Hence $S = 3$.

28. (i) $\frac{4}{3}$; (ii) 3.

29. (i) $\frac{5}{2}$; (ii) $\frac{7}{2}$.

30. If $S = a + ar + ar^2 + \cdots$, then $rS = ar + ar^2 + \cdots$. Hence $S - rS = a$. This give $S = \frac{a}{1-r}$. Obviously $r \neq 1$ because dividing by zero is the most heinous of mathematical crimes. Anyway, if $r = 1$, then $rS = S$ and so subtracting rS from S is fairly useless. Clearly if $r = 1$, then S is infinite.

    There's something screwy too if $r > 1$. What's going on here? So $S = \frac{a}{1-r}$ provided $r < 1$.

Oh dear! what if r is negative? It turns out that $S = \frac{a}{1-r}$ provided $|r| < 1$.

Check out the previous exercise to see that this result holds.

31. (i) $1 + 2x + 4x^2 + 8x^3 + 16x^4 + \cdots$; (ii) $1 - x + x^2 - x^3 + x^4 \cdots$; (iii) $1 + x^2 + x^4 + x^6 + x^8 \cdots$.

32. $(1-x)^{-1}(1+x)^{-1} = (1+x+x^2+\cdots)(1-x+x^2\cdots) = 1-x^2+x^4-\cdots$.

$$\frac{1}{1-x^2} = \frac{1/2}{1-x} + \frac{1/2}{1+x} = \frac{1}{2}(1 + x + x^2 + \cdots) + \frac{1}{2}(1 - x + x^2\cdots)$$
$$= 1 - x^2 + x^4 - \cdots.$$

33. They should.

34. (i) $1 - 2x + 3x^2 - 4x^3 + \cdots$; (ii) $1 - 3x + 6x^2 - 10x^3 + \cdots$; (iii) $1 - 4x + 10x^2 - 20x^3 + \cdots$.

35. $\frac{d}{dx}(1+x)^{-1} = \frac{-1}{(1+x)^{-2}}$. The derivative of $1 - x + x^2 - x^3 + x^4 - \cdots$ is $-1 + 2x - 3x^2 + 4x^3 \cdots$. So $\frac{1}{(1+x)^{-2}} = 1 - 2x + 3x^2 - 4x^3 + \cdots$.

Similar things happen for $\frac{1}{(1+x)^3} = \frac{-1}{2}[\frac{d}{dx}(1+x)^{-2}]$ and $(1+x)^{-4}$.

36.   (i) $1 + 2x^2 + 3x^4 + 4x^6 + \cdots$;

   (ii) $x + x^2 + x^3 + x^4 + \cdots$;

   (iii) $\dfrac{1}{x(1-x)} = \dfrac{1}{x} + \dfrac{1}{1-x} = \dfrac{1}{x} + 1 + x + x^2 + x^3 + \cdots$;

   (iv) $1 - 3x^2 + 6x^4 - 10x^6 + \cdots$;

   (v) $\dfrac{1}{2-x} = \dfrac{1}{2(1-x/2)} = \dfrac{1}{2}[1 + \dfrac{x}{2} + \dfrac{x^2}{4} + \dfrac{x^3}{8} + \cdots]$;

   (vi) $\dfrac{1}{4-x^2} = \dfrac{1}{4}[1 + \dfrac{x^2}{4} + \dfrac{x^4}{16} + \dfrac{x^6}{64} + \cdots]$;

   (vii) $1 + 4x^2 + 16x^4 + 64x^6 + \cdots$;

   (viii) $1 + 36x^2 + 810x^4 + 14580x^6 + \cdots$;

   (ix) $\dfrac{1}{(1+x)(1+2x)} = \dfrac{-1}{1+x} + \dfrac{2}{1+2x} = -[1 - x + x^2 - x^3 + x^4 \cdots] + 2[1 - 2x + 4x^2 - 8x^3 + 16x^4 - \cdots] = 1 - 3x + 7x^2 - 15x^3 + 31x^4 \cdots$.

37. $a_{10} = 55$ but those others take one heck of a lot of calculating. Perhaps a little programming of the old computer is in order.

(Warning: $a_{100} \cong 3.54 \times 10^{20}$. Is your computer accurate enough for this?)

38. $1, 2, 3, 5, 8, 13, 21, 34$. (That was easier!)

39. Since 1 is even and 2 is odd, then the third term has to be odd. But the fourth term is the sum of an even and an odd, and so is odd. Odd + odd is even. We, that is, you, should now be able to prove that the terms go odd, even, odd; odd, even, odd; for ever.

Hence, no consecutive evens but lots of consecutive odds.

40. Isn't the sequence odd, odd, even? In that case a_{3k} is even for all $k \in \mathbb{N}$. (A little bit of mathematical induction should give this.)

41. (i) 3, 6, 12, 24, 48; (ii) 1, 4, 16, 64, 256; (iii) 3, 8, 13, 18, 23; (iv) 1, −3, 5, 21, 437; (v) 0, 1, 1, 2, 3; (vi) 1, 1, 3, 7, 17; (vii) 1, 1, 0, −1, −1; (viii) 1, 1, 0, 1, −1.

42. (i) $a_n = 2a_{n-1}, a_1 = 5$;

   (ii) $a_n = 3a_{n-1}, a_1 = 5$;

   (iii) $a_n = a_{n-1} + 2, a_1 = 5$;

   (iv) $a_n = a_{n-1} + a_{n-2}, a_1 = 5, a_2 = 7$;

   (v) $a_n = 2a_{n-2} + a_{n-1}, a_1 = 5, a_2 = 7$;

   (vi) $a_n = 2(a_{n-1} + a_{n-2}), a_1 = 5, a_2 = 7$.

43. (i) $a_n = 2^n$; (ii) $a_n = 3^{n-1}$; (iii) $a_n = 3 \times 4^{n-1}$; (iv) $a_n = -5^{n-1}$.

44. I will only show Step 3 of each of the proofs by mathematical induction.

   (i) assume $a_k = 2^k$. Then $a_{k+1} = 2a_k = 2 \times 2^k = 2^{k+1}$;
   (ii) assume $a_k = 3^{k-1}$. Then $a_{k+1} = 3a_k = 3 \times 3^{k-1} = 3^k$;
   (iii) assume $a_k = 3 \times 4^{k-1}$. Then $a_{k+1} = 4a_k = 4 \times 3 \times 4^{k-1} = 3 \times 4^k$;
   (iv) assume $a_k = -5^{k-1}$. Then $a_{k+1} = 5a_k = 5(-5^{k-1}) = -5^k$.

45. I will show a few terms of the sequence, then my guess. I'll leave the induction to you.

   (i) $a_1 = 5, a_2 = 10, a_3 = 20, a_4 = 40$. Guess $a_n = 5 \times 2^{n-1}$;

   (ii) $a_1 = 5, a_2 = 8, a_3 = 11, a_4 = 14$. Guess $a_n = 5 + 3(n-1)$;

   (iii) $a_1 = 5, a_2 = 10, a_3 = 15, a_4 = 20$. Guess $a_n = 5n$;

   (iv) $a_1 = 5, a_2 = 3, a_3 = 1, a_4 = -1$. Guess $a_n = 5 - 2(n-1)$;

   (v) $a_1 = 5, a_2 = 11, a_3 = 23, a_4 = 47$. Guess $a_n = 5 \times 2^{n-1} + 2^{n-2} + 2^{n-3} + \cdots + 2 + 1 = 5 \times 2^{n-1} + 2^{n-1} - 1 = 6 \times 2^{n-1} - 1$;

   (vi) $a_1 = 4, a_2 = 10, a_3 = 28, a_4 = 82$. Guess $a_n = 3^n + 1$.

46. Since $a_n = 2a_{n-1}$ and $a_1 = 2$, we can use Exercise 43(i) to give $a_n = 2^n$. Hence there are 2^n subsets of a set with n elements.

47. $a_1 = 4, a_2 = 16$. Suppose we try to make up an n-digit number from an $(n-1)$-digit number. We can do this by placing 1, 2, 3 or 4 in front of the $(n-1)$-digit number. There are therefore four times as many n-digit numbers as there are $(n-1)$-digit numbers. Hence $a_n = 4a_{n-1}$. Since $a_1 = 4$, try the guess $a_n = 4^n$. (Prove this using mathematical induction.)

48. $a_1 = 5, a_2 = 25$ and $a_n = 5a_{n-1}$ (using the same sort of argument as in the last exercise). This should give $a_n = 5^n$.

49. $b_1 = 1$ (just the number 1 is allowed), $b_2 = 6$ (here we have 12, 13, 14, 21, 31, 41). Suppose we now look at an n-digit number. If this number starts with a 2, 3, or 4, then it's followed by $(n - 1)$-digit where 1 occurs an odd number of times. There are b_{n-1} such possible $(n - 1)$-digit numbers. If we add 2, 3, or 4 to the left of these we get $3b_{n-1}$ n-digit numbers with an odd number of 1's.

But our n-digit number may begin with a 1.

If it does, then there are an even number of 1's from there on. There are $4^{n-1} - b_{n-1}$ such possible $(n-1)$-digit numbers. This is all possible $(n - 1)$-digit numbers made from 1, 2, 3, 4 minus those with an odd number of 1's.

Combining the two possible starts, we see that
$b_n = 3b_{n-1} + (4^{n-1} - b_{n-1})$.
Hence $b_n = 4^{n-1} + 2b_{n-1}$.

Guess $b_n = \frac{1}{2}(4^n - 2^n)$. Prove this by mathematical induction.

Now $c_n = 4^n - b_n$. This was actually established earlier. So
$c_n = \frac{1}{2}(4^n + 2^n)$.

50. (i) $\alpha = 1, \beta = 3; K = 1, L = 0; \therefore a_n = 1$;

(ii) $\alpha = 1, \beta = 3; K = 0, L = 1; \therefore a_n = 3^n$;

(iii) $\alpha = 1, \beta = 3; K = -3, L = \frac{3}{4}; \therefore a_n = 4 \times 3^{n-1} - 3$;

(iv) $\alpha = 2, \beta = 3; K = -\frac{1}{2}, L = \frac{1}{3}; \therefore a_n = 3^{n-1} - 2^{n-1}$;

(v) $\alpha = \beta = 2; K = -\frac{1}{4}, L = \frac{1}{4}; \therefore a_n = 2^{n-2}(n - 1)$.

51. $x^2 - x - 1 = 0$ gives $\alpha = \frac{1}{2}(1 + \sqrt{5}), \beta = \frac{1}{2}(1 - \sqrt{5}), K = \frac{1}{\sqrt{5}}, L = -\frac{1}{\sqrt{5}}$;
$\therefore a_n = \frac{1}{2^n \sqrt{5}}[(1 + \sqrt{5})^n - (1 - \sqrt{5})^n]$. (Check this out for a few terms. Looks suspiciously complicated doesn't it?)

52. $\{1, 2\}$ gives $a_1 = 2$; $\{11, 12, 21\}$ gives $a_2 = 3$; $\{111, 112, 121, 211, 212\}$ gives $a_3 = 5$.

Suppose we make an n-digit number with no consecutive 2's. Then it either starts with a 1 or a 2. If it starts with a 1 we can add any $(n - 1)$-digit number with no consecutive 2's. There are a_{n-1} of these. If it starts with a 2, then the next number must be a 1 because there are no consecutive 2's. But after this 1 we can put any $(n - 2)$-digit number. There are therefore a_{n-2} n-digit numbers with no consecutive 2's which start with 21. Hence $a_n = a_{n-1} + a_{n-2}$. (The theorem shows us how to find the n-th term of this sequence.)

53. Isn't this the same as the last exercise?

54. There are a_{n-1} such numbers which start with a 1. There are $2a_{n-2}$ such numbers that start with a 2 (they actually start with 21, 23). There are $2a_{n-2}$ such numbers which start with a 3. Hence $a_n = a_{n-1} + 4a_{n-2}$.

55. There are a_{n-1} which start with B and another a_{n-1} which start with C. There are $2a_{n-2}$ which start with A (they start AB or AC). Hence
$$a_n = 2a_{n-1} + 2a_{n-2}.$$
Clearly $a_1 = 3$ and $a_2 = 8$.

Using the theorem, $x^2 - 2x - 2 = 0$ gives $\alpha = 1 + \sqrt{3}$, $\beta = 1 - \sqrt{3}$. Then $K = \frac{1}{6}(3 + 2\sqrt{3})$, $L = \frac{1}{6}(3 - 2\sqrt{3})$. Hence
$$a_n = \frac{1}{6}[(3 + 2\sqrt{3})(1 + \sqrt{3})^n + (3 - 2\sqrt{3})(1 - \sqrt{3})^n].$$

56. Let's try Step 3 in the induction, where $\alpha \neq \beta$. So we may assume that $a_k = K\alpha^k + L\beta^k$. Then
$$a_{k+1} = Aa_k + Ba_{k-1} = A(K\alpha^k + L\beta^k) + B(K\alpha^{k-1} + L\beta^{k-1})$$
$$= (A\alpha + B)K\alpha^{k-1} + (A\beta + B)\beta^{k-1}.$$

So all we've got to do is to show that $A\alpha + B = \alpha^2$ and $A\beta + B = \beta^2$ and we are done. How to do that? (See after Exercise 19 (revisited).) So, in the meantime, let's try Step 3, when $\alpha = \beta$.

We can assume that $a_k = (K + kL)\alpha^k$. Hence
$$a_{k+1} = Aa_k + Ba_{k-1} = A[K + kL]\alpha^k + B[K + (k-1)L]\alpha^{k-1}$$
$$= \{[A\alpha + K]K + [Ak\alpha + B(k-1)]L\}\alpha^{k-1}.$$

We now have to show that $A\alpha + K = \alpha^2$ and $Ak\alpha + B(k-1) = (k+1)\alpha^2$. At least the former requirement is the same as in the $\alpha \neq \beta$ case!

57. Using the Theorem we get $a_n = \frac{1}{2\sqrt{2}}[(1 + \sqrt{2})^n - (1 - \sqrt{2})^n]$.

The Binomial Theorem expansion for $(1 + \sqrt{2})^n$ is $\sum_{r=0}^{n} {}^nC_r(\sqrt{2})^r$. Hence $a_n = \frac{1}{2\sqrt{2}}\left[\sum_{r=0}^{n} {}^nC_r(\sqrt{2}^r) - \sum_{r=0}^{n} {}^nC_r(-1)^r(\sqrt{2}^r)\right]$. If r is even, the terms ${}^nC_r(\sqrt{2})^r$ and ${}^nC_r(-1)^r(\sqrt{2})^r$ cancel. If r is odd, they give $2{}^nC_r(\sqrt{2})^r$. So

$$a_n = \frac{2}{2\sqrt{2}}[{}^nC_1\sqrt{2} + {}^nC_3(\sqrt{2})^3 + {}^nC_5(\sqrt{2})^5 + \cdots]$$
$$= {}^nC_1 + 2{}^nC_3 + 4{}^nC_5 + 8{}^nC_7 + \cdots$$
$$= n + \sum_{p \geq 1} 2^p {}^nC_{2p+1}$$

where the summation keeps going up to the last odd integer less than or equal to n.

Let $n = 2^k m$, where m is odd. Then $a_n = 2^k m + \sum_{p \geq 1} 2^p {}^nC_{2p+1}$. We therefore have to show that the summation is of the form $2^{k+1}m$ for some integer m. (Why?)

This is all done later (57 revisited). We just note here that this problem was submitted by Bulgaria for the 29th IMO in Canberra.

58.   (i) $f(x) = x + 2x^2 + 3x^3 + 4x^4 + \cdots = x(1-x)^{-2}$;

(ii) $f(x) = x + x^3 + x^5 + \cdots = x(1-x^2)^{-1}$;

(iii) $f(x) = x - x^2 + x^3 - x^4 + \cdots = x(1+x)^{-1}$;

(iv) $f(x) = x + 3x^2 + 6x^3 + 10x^4 + 15x^5 + 21x^6 + \cdots = x(1-x)^{-3}$;

(v) $f(x) = 2x + 4x^2 + 8x^3 + 16x^4 + \cdots = 2x(1-2x)^{-1}$;

(vi) $f(x) = 2x - x^2 + 1/2x^3 - 1/4x^4 \cdots = 4x(2+x)^{-1}$.

59.   (i) $x(1-3x)^{-1}$; $a_n = 3^{n-1}$;

(ii) $3x(1-2x)^{-1}$; $a_n = 3 \times 2^{n-1}$;

(iii) $x(1-4x)^{-1}$; $a_n = 4^{n-1}$;

(iv) $x(1+3x)^{-1}$; $a_n = (-3)^{n-1}$;

(v) $x(1-x)^{-2}$; $a_n = n$;

(vi) $2x(1-x)^{-2}$; $a_n = 2n$.

60.   (i) $x(1-5x+6x^2)^{-1}$;

(ii) $(4x - 3x^2)(1 - 2x - 3x^2)^{-1}$;

(iii) $2x(1-3x)(1-4x-5x^2)^{-1}$;

(iv) $x^2(1-x-x^2)^{-1}$.

I will show you how to find a_n later (see Example 5).

61. $f(x) = \sum_{n=1}^{\infty} a_n x^n = a_1 x + a_2 x^2 + \sum_{n=3}^{\infty} (A a_{n-1} + B a_{n-2}) x^n$

$$= a_1 x + a_2 x^2 + Ax \sum_{n=3}^{\infty} a_{n-1} x^{n-1} + Bx^2 \sum_{n=3}^{\infty} a_{n-2} x^{n-2}$$

$$= a_1 x + a_2 x^2 + Ax(f(x) - a_1 x) + Bx^2 f(x).$$

Hence $f(x)[1 - Ax - Bx^2] = a_1 x + x^2 (a_2 - Aa_1)$. The result then follows.

62. See 60 (revisited) towards the end of the solutions.

63.   (i) $(3^{n-1} - 2^{n-1})$;   (ii) $\dfrac{1}{4}[3^n - (-1)^n]$;

(iii) $\dfrac{1}{6}[5^n - (-1)^n]$;   (iv) $\dfrac{1}{15}[2 \times 5^n - 5(-1)^n]$;

(v) $2^{n+1} - 3$;   (vi) $3^n - 1$.

64. Let $1 - x - x^2 = (1 - \alpha x)(1 - \beta x)$.

Then $F(x) = \dfrac{1}{(\alpha - \beta)} \left[\dfrac{1}{1 - \alpha x} - \dfrac{1}{1 - \beta x} \right]$

$$= \dfrac{1}{(\alpha - \beta)} \{ \sum_{n=0}^{\infty} (\alpha x)^n - \sum_{n=0}^{\infty} (\beta x)^n \}$$

$$= \dfrac{1}{(\alpha - \beta)} \sum_{n=0}^{\infty} (\alpha^n - \beta^n) x^n.$$

Hence $a_n = \frac{1}{(\alpha-\beta)} \sum_{n=0}^{\infty} (\alpha^n - \beta^n)$. (When $n = 0$, $\alpha^n - \beta^n = 0$.)

But what are α and β? See later (63 revisited).

65. If the first square is red or blue, there are no two adjacent white squares in the remaining $n - 1$ positions. Hence there are $2a_{n-1}$ boards which start as red or blue.

   If the first square is white, the next square is red or blue and these contribute $2a_{n-2}$ to the required number.

   Hence $a_n = 2a_{n-1} + 2a_{n-2}$. It is easy to see that $a_1 = 3$ and $a_2 = 8$. This recurrence relation can now be solved using generating functions. Do you get the same answer as Exercise 55?

66. The generating function method is clearly longer, though it is a method that will work for a wider range of recurrence relations than the theorem. However, there are theorems to deal with more general situations.

   How could the theorem be changed to deal with
   $a_n = Aa_{n-1} + Ba_{n-2} + Ca_{n-3}$? (See 66 revisited.)

67. (a) This can easily be done by induction. (See later 67 revisited.)

   (b) Let $i = (k - 1)n$ and $j = n$. Then from (a) we have
   $a_{kn} = a_{(k-1)n}a_{n-1} + a_{(k-1)n+1}a_n$.
   This starts off another mathematical induction proof.

   (c) Suppose a_n and a_{n+1} are both divisible by t. Since
   $a_{n+1} = a_n + a_{n-1}$,
   a_{n-1} is also divisible by t. Repeating this process we see that a_{n-2} is divisible by t. Eventually we see that a_2 and a_1 are divisible by t. Hence $t = 1$.

   There must be a way of doing this without using mathematical induction. (See 67 revisited.)

68. By Exercise 67(a), $a_{2000} = a_{1990}a_9 + a_{1991}a_{10}$. Since a_{1990} and a_{1991} are relatively prime (Exercise 67(c)) then any divisor of a_{2000} and a_{1990} also divides a_{10}. But by Exercise 67(b), a_{10} divides a_{1990} and a_{2000}. Hence a_{10} is the greatest common divisor of a_{1990} and a_{2000}.

   You might like to show that if $(m, n) = s$, then $(a_m, a_n) = a_s$, where (p, q) means the highest common factor of p and q.

   (This question is based on one submitted by the Republic of Korea at the 29th IMO.)

69. Have a look at the MacTutor web site.

   It's clearly not a good model — nothing died.

70. See the MacTutor site.

71. (a) $p_1 = 0$, $p_2 = 1$, $p_3 = 2$, $p_4 = 9$;
    (b) Well?
    (c) $p_5 = 44$, $p_6 = 265$. Yes.
72. What did you get? (Not very far if you tried constants for A and B.)
73. (a) $p_4 = 9$. With n houses if 1 and 5 are swapped, there are p_{n-2} possible bad deliveries.
    (b) $5p_4 = 9$. In general, $(n-1)p_{n-2}$. There are $n-1$ possible letters to swap with.
74. (a) p_5. p_{n-1};   (b) $5p_5$. $(n-1)p_{n-1}$.
75. $p_n = (n-1)p_{n-1} + (n-1)p_{n-2}$. To swap or not to swap, that is the question. The generalisation comes immediately from looking at the special cases.
76. What did you get?
77. Looks like mathematical induction on
    $$p_n = n!\left(1 - \tfrac{1}{1!} + \tfrac{1}{2!} - \tfrac{1}{3!} + \tfrac{1}{4!} - \cdots + \tfrac{(-1)^n}{n!}\right).$$
    Alternatively we could use inclusion-exclusion. But that's another chapter (Chapter 6).
19. (revisited) Since $2na_n = (2n-3)a_{n-1}$ we have
    $-a_{n-1} = 2na_n - 2(n-1)a_{n-1}$.
    Hence $a_{n-1} = 2(n-1)a_{n-1} - 2na_n$.
    For any k then $a_k = 2k\,a_k - 2(k+1)a_{k+1}$.
    Hence
    $$\begin{aligned}
    \sum_{k=1}^{n} a_k &= \sum_{k=1}^{n} 2k\,a_k - 2(k+1)a_{k+1} \\
    &= (2a_1 - 2 \times 2a_2) + (2 \times 2a_2 - 2 \times 3a_3) \\
    &\quad + (2 \times 3a_3 - 2 \times 4a_4) + \cdots + [2na_n - 2(n+1)a_{n+1}] \\
    &= 2a_1 - 2(n+1)a_{n+1} \\
    &= 1 - 2(n+1)a_{n+1}.
    \end{aligned}$$

    Since $a_{n+1} > 0$, $\sum_{k=1}^{n} a_k < 1$.
    [If $a_1 = \tfrac{1}{2}$ and $a_n = \left(\tfrac{pn-q}{pn}\right)a_{n+1}$, for $n \geq 1$, what can be said about $\sum_{k=1}^{n} a_k$ and when?]
56. (revisited) $A\alpha + B = \alpha^2$ since α is a root of $Ax + B = x^2$. Similarly $A\beta + B = \beta^2$.
    But what about $Ak\alpha + B(k-1) = (k+1)\alpha^2$?
    Well $Ak\alpha + B(k-1) = Ak\alpha + Bk - B = k(A\alpha + B) - B = k\alpha^2 - B$.
    All we want to show now is that $-B = \alpha^2$. Is this likely? (See later.)
57. (revisited) $2^p\,{}^nC_{2p+1} = 2^p n \dfrac{(n-1)\cdots(n-2p)}{(2p+1)!} = 2^{k+p}\dfrac{m}{2p+1}{}^{n-1}C_{2p}$.

Now $2^p \, {}^nC_{2p+1}$ is an integer and $2p + 1$ isn't a factor of 2^p. Hence $\frac{m}{2p+1} \, {}^{n-1}C_{2p}$ is an integer. This means that $2^p \, {}^nC_{2p+1}$ is divisible by 2^{k+p}. So $a_n = n + \sum_{p \geq 1} 2^p \, {}^nC_{2p+1} = 2^k m + 2^{k+1} M$, for some integer M. The result now follows.

60. (revisited) (i) This is done in Example 6;

(ii) Since
$$(1 - 2x - 3x^2)^{-1} = (1 - 3x)^{-1}(1 + x)^{-1} = \tfrac{3}{4}(1 - 3x)^{-1} + \tfrac{1}{4}(1 + x)^{-1},$$
we have
$$(4x - 3x^2)(1 - 2x - 3x^2)^{-1} = \sum_{n=1}^{\infty} \tfrac{1}{4}[(-1)^{n-1}7 + 3^{n+1}]x^n;$$

(iii) $(2x - 6x^2)\left[\frac{5/6}{1-5x} + \frac{1/6}{1+x}\right] = \sum_{n=1}^{\infty} \tfrac{1}{6}[4 \times 5^{n-1} + 8(-1)^{n-1}]x^n;$

(iv) $x^2(1 - x - x^2)^{-1} = xF(x) = \frac{x}{(\alpha-\beta)}\left[\sum_{n=1}^{\infty}(\alpha^n - \beta^n)x^n\right]$. (See Exercise 64 for full details.)

64. (revisited) But $1 - x - x^2 = (1 - \alpha x)(1 - \beta x) = 1 - (\alpha + \beta)x + \alpha\beta x$. Hence $\alpha\beta = -1$ and $\alpha + \beta = 1$. So $\alpha - \frac{1}{\alpha} = 1$. This gives $\alpha^2 - \alpha - 1 = 0$. Therefore $\alpha = \frac{1 \pm \sqrt{5}}{2}$ and $\beta = \frac{1 \pm \sqrt{5}}{2}$.

Without loss of generality, choose $\alpha = \frac{1+\sqrt{5}}{2}$ and $\beta = \frac{1-\sqrt{5}}{2}$. Hence $a_n = \frac{1}{\sqrt{5}}\left[\left(\frac{1+\sqrt{5}}{2}\right)^n - \left(\frac{1-\sqrt{5}}{2}\right)^n\right]$, which is what we got in Exercise 51. (Would we get the same result if we had chosen $\alpha = \frac{1-\sqrt{5}}{2}$ and $\beta = \frac{1+\sqrt{5}}{2}$? Why?)

66. (revisited) Try $x^3 = Ax^2 + Bx + C$.

67. (revisited) (a) The crucial step is
$$a_{i+j+1} = a_{i+j} + a_{i+j-1} = (a_i a_{j-1} + a_{i+1}a_j) + (a_i a_{j-2} + a_{i+1}a_j).$$
Alternatively: (a) Use the fact that $a_n = \frac{\alpha^n - \beta^n}{\alpha - \beta}$.

(b) $x^n - y^n = (x - y)(x^{n-1}y + \cdots + y^{n-1})$, so let $x = \alpha^k$ and $y = \beta^k$.

(c) Can this be done this way?

56. (revisited again) We want to show that $-B = \alpha^2$. Well, remember that $x^2 - Ax - B = 0$ and that $(x - \alpha)^2 = 0$. Hence $x^2 - 2\alpha x + \alpha^2 = 0$. Since this last quadratic is precisely the same as $x^2 - Ax - B = 0$, $A = 2\alpha$ but, more importantly, $-B = \alpha^2$.

Chapter 2

Geometry 3

In this chapter I want to move on from the two geometry chapters in *First Step* and look at several things relating to circles, specifically circles that are related to triangles. So I talk about angles in a circle, the Sine Rule, incircles and excircles, as well as circumcircles. In the process I need to think about altitudes of triangles and where they meet — at the orthocentre of the triangle. Similarly I'll look at the centroid of a triangle where the medians intersect. This will lead us on to thinking about the Euler line of a triangle and the big theorem of the chapter — the Nine Point Circle Theorem.

Because of the great number of new concepts defined here, I have a Glossary at the back of this chapter in Section 2.8. That should enable you to quickly find out what something is when the need arises.

In all geometry problems it is useful to draw a diagram. The bigger the diagram the better. Don't scrimp on space. It is much easier to put all the points you need on a big diagram. And it is much harder to be confused or get led down wrong paths of intuition this way.

As some geometry problems get a bit sticky at times, I've added a "Hints" section in this chapter. Your aim should be to solve the exercises without recourse to the hints but at times this may prove difficult. At the stage when you're just about to succumb and look for a hint, ask some of your friends for their ideas. Maybe a passing teacher will provide some fresh insight. Even telling fathers who are cooking breakfast, despite the fact they haven't got a clue what you're talking about, may provide you with inspiration. It's strange how speaking your problems out loud to someone else will often get you started.

Oh! And I know I seem to set a great store by *proving* things. Don't worry about that too much the first time you go through this chapter. Try to see *why* things are probably so and get a good overall picture. Then go back and wrestle with the proof later. Unusually, the answers are in the penultimate section of the chapter. However, there is frequently more than one proof/solution to a geometry problem so it is more than possible that you will get a correct answer that is different to mine. Again, check your proof with a friend to see what they think of it.

2.1. The Circumcircle

Let's see what you've remembered from Chapter 5 of *First Step*. Suppose I've got a triangle and I want to construct a circle which passes through each vertex. How do I go about it? How do I find the centre?

The circle which goes through the three vertices of a triangle is called the triangle's *circumcircle*. The centre of this circle, not surprisingly, is called the *circumcentre* of the triangle. Its radius is the *circumradius*.

Exercises

1. (a) Where is the circumcentre of a right angled triangle?
   (b) Where is the circumcentre of an equilateral triangle?
   (c) Where is the circumcentre of an isosceles triangle?
2. If the circumcentre of a triangle lies on one of the sides of the triangle, what kind of triangle is it?
3. For what kinds of isosceles triangles does the circumcircle lie *inside* the triangle?
4. (a) What is the circumradius of a right angled triangle?
   (b) What is the circumradius of an equilateral triangle?
5. Draw the circumcircle around the right triangle ABC, where AC is the hypotenuse. Let B' be any point on the semi-circle from A to C which contains B. What is the angle $AB'C$? Use ruler, compass and protractor if you have to but some geometry software might be better.
6. Draw the circumcircle around the equilateral triangle ABC. Let C' be any point on the arc of the circle between A and B which contains C. What is $\angle AC'B$?
7. Generalise Exercises 5 and 6.

As a result of the last few exercises it's beginning to look as if the angle subtended at the circumference of a circle by a fixed chord is constant. The angles AC_iB in Figure 2.1 are *subtended* by the chord AB being the angles formed by the lines from C_i to the ends of the chord AB. Is it true that the angles AC_iB are all the same for $i = 1, 2, 3, 4$?

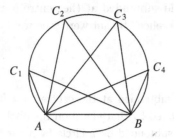

Figure 2.1.

Try drawing some circles, electronically or otherwise. Take any chord AB and see if all the angles ACB are the same, where C is any point on a segment of the circle.

If this is true, of course, there's got to be a reason for it. How could we possibly show that all those angles are the same? What is constant about the circle? Certainly AB is. What if we looked at the isosceles triangle with base AB and vertex C on the circle? What about looking at the triangle AOB, where O is the circumcentre of the triangle? (See Figure 2.2.)

How does that help?

Well, $\triangle AOC$ is an isosceles triangle as $AO = OC$ (since they are radii), so $\angle OAC = \angle OCA$. Hmm. But $\angle DOA = \angle OAC + \angle OCA$ (because $\angle DOA$ is an external angle opposite the two internal angles $\angle OAC$ and $\angle AOC$). So $\angle DOA = 2\angle OCA$. But that means that $\angle DOB = 2\angle OCB$ too. So $\angle AOB = 2\angle ACB$.

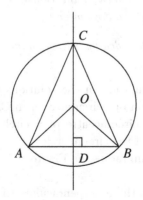

Figure 2.2.

So what? Is the angle subtended at the centre always twice the angle subtended at the circumference? If that were true we'd be finished and with a very pretty result.

Exercises

8. Show that the angle subtended at the centre by a fixed chord is **always** twice the angle subtended by that chord on the circle itself.
9. Prove that the angle subtended on a circle by a fixed chord is a constant size.

   Do two different chords of equal length subtend the same angle on the same circle?

10.
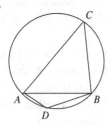
Is $\angle ADB = \angle ACB$?
If so, prove it. If not, find a relationship between the two angles and prove that the relationship actually holds.

11. What angle is subtended at the circumference by a chord which is a diameter?
12. A **cyclic quadrilateral** is a four-sided figure whose four vertices lie on the circumference of a circle. Let $ABCD$ be a cyclic quadrilateral. What relation holds between opposite angles in a cyclic quadrilateral?

    For instance, is $\angle ABC = \angle ADC$?

Theorem 1. *The angle subtended at the centre of a circle by a chord is twice the angle subtended on the circumference.*

Theorem 2. *All angles subtended by a fixed chord in the same segment of a circle are equal.*

The proof of these results are in the solutions to Exercises 8 and 9. But I've wandered away from our circumcircle. In Chapter 5 of *First Step* we found how to find the circumcentre of a circle, but how big is the **circumradius**, the radius of the circumcircle?

Suppose R is the circumradius in Figure 2.3. Then $OA = OB = OC = R$.

If OL, OM, ON are the three perpendiculars to the sides, then $AL = LB, BM = MC, CN = NA$.

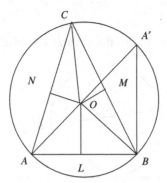

Figure 2.3.

If we extend the line segment AO till it hits the circle again at A', then AA' is a diameter of the circumcircle. What can we say about the triangle ABA'? What can we say about $\angle AA'B$? Has it got anything to do with any of the angles of the original triangle?

What is $\frac{AB}{AA'}$?

Exercises

13. Prove that in Figure 2.3, $AL = LB$.
14. Show that $\frac{c}{\sin C} = 2R$, where c is the length of the side AB and $C = \angle ABC$.
15. What is $\frac{b}{\sin B}$ equal to, where b is the length of side AC and $B = \angle ABC$?
16. Repeat the last exercise with $\frac{a}{\sin A}$.
17. Find an equation linking Δ, the area of any triangle ABC, a, b, c the length of the sides of the triangle and R, the circumradius. Test out your conjecture on the triangles of Exercise 4. Prove that the equation you come up with holds for any triangle.
18. In Figure 2.3, the circumcentre lies **inside** $\triangle ABC$. Do the results of Exercises 14–16 still hold if the circumcentre lies **outside** $\triangle ABC$?

Exercises 14–16 and 18, have proved the Sine Rule for you. This is, in *any* triangle ABC,

$$\frac{a}{\sin A} = \frac{b}{\sin B} = \frac{c}{\sin C}.$$

In addition, we know that $\frac{a}{\sin A}$ is actually equal to twice the circumradius of $\triangle ABC$.

Exercises

19. Let AX be the tangent at A to the circumcircle of $\triangle ABC$ (see diagram).

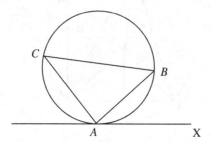

(a) If $\angle XAC = 90°$, show that $\angle XAB = \angle ACB$.
(b) Show that $\angle XAB = \angle ACB$ no matter what size $\angle XAC$ is.

20. Let p, q be the radii of two circles which have one point, A, in common and which have BC as a tangent, with the circles touching BC at B and C, respectively. Show that $pq = R^2$, where R is the circumradius of $\triangle ABC$.

2.2. Incircles

The biggest circle you can fit **inside** a triangle is called its **incircle**. Obviously, the centre of the incircle is the **incentre**, denoted here by O. And the radius of the incircle is the **inradius**. The incircle and incentre of a triangle are shown in Figure 2.4.

Some of the questions to ask now are: How can you construct an incircle? Where is the incentre? What is the inradius, r?

Some of these questions were answered in Chapter 5 of *First Step* but let's delve a little deeper into them here.

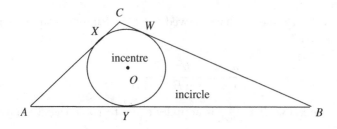

Figure 2.4.

Exercises

21. Draw the two tangents from the point D external to a circle, centre O. Suppose the tangents touch the circle at E and F. What is the relation between the lengths of the line segments DE and DF? What is the relation between $\angle ODE$ and $\angle ODF$?

22. Show how to find the incentre of $\triangle ABC$.
    Construct the incircle of $\triangle ABC$ using ruler and compass.

23. In Exercise 17 we found an equation linking the area of the $\triangle ABC$ and R, the circumradius. Can you find an equation which links the area of $\triangle ABC$ and r, the inradius?

24. What is the size of the inradius for the triangle with sides $3, 4, 5$?

25. Is R ever equal to $\frac{5}{2}r$? Is R ever equal to $2r$?

In Exercise 22 we needed to look at the bisectors of the angles of a triangle. Why not look at the bisectors of the **sides** of a triangle? A line from a vertex of a triangle to the midpoint of the opposite side, is called a **median**. In Figure 2.5, AL, BM, CN are all medians, so $AN = NB, BL = LC, CM = MA$. Figure 2.5 rather seems to suggest that the three medians of a triangle are concurrent.

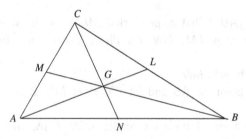

Figure 2.5.

To see whether this is true or not look at Figure 2.6.

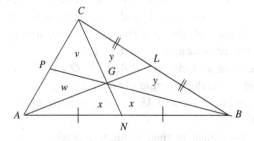

Figure 2.6.

Now suppose that just the medians AL and CN meet at G. Draw BG produced to intersect AC at P. Let $CP = \alpha AP$. Clearly we'd love to show that $\alpha = 1$.

First note that area $\triangle AGN =$ area $\triangle BGN$. After all, these triangles have the same base and the same height. Similarly area $\triangle BGL =$ area $\triangle CGL$.

But it's also true (for the same reason) that area $\triangle CAN =$ area $\triangle BCN$. So $v + w + x = 2y + x$. In other words, $v + w = 2y$.

Now we can repeat this trick on $\triangle$'s BAL and CAL to give $2x + y = v + w + y$. So $v + w = 2x$. We've now established that $x = y$.

If we now look at $\triangle$'s CGP and AGP we see that $v = \alpha w$ because the triangles have the same height but the base of one is α times the length of the base of the other.

Looking at $\triangle$'s CBP and ABP we find that $v + 2y = \alpha(w + 2x)$. So $2y = \alpha 2x$. However, we know that $x = y$, so $\alpha = 1$. This proves just what we wanted.

The point G where the three medians meet, is called the **centroid** of $\triangle ABC$.

Exercises

26. Prove that $ML \| AB$. (That is, prove that ML is parallel to AB.)
27. Show that the lines LM, MN, NL divide $\triangle ABC$ into four congruent triangles.
28. Prove that AL bisects MN.
29. Let Y be any point on AB and let CY meet MN at X. Find the ratio $CX : CY$.
30. Show that the areas of the six triangles AGN, BGN, BGL, CGL, CGM, AGM of Figure 2.5, are all equal.
31. Use areas to find the ratios $\frac{AG}{GL}$, $\frac{BG}{GM}$ and $\frac{CG}{GN}$.
32. Another set of three lines in a triangle which are concurrent, are the **altitudes** of the triangles. These are the line segments drawn from a vertex to the opposite side of the triangle and perpendicular to the opposite side. Show that the three altitudes of a triangle are concurrent. This common point is called the **orthocentre** of the triangle.
33. The feet of the three altitudes of a $\triangle ABC$, P, Q and R, are the vertices of a triangle called the **orthic triangle** of triangle ABC. Show that the orthocentre of $\triangle ABC$ is the incentre of its orthic triangle, that is, $\triangle PQR$.
34. What triangles are similar to their orthic triangles?

35. (a) Describe the orthic triangle of a right angled triangle.
    (b) Can an orthic triangle have a right angle?

2.3. Excircles

We know that three non-collinear points define a unique circle. But it occurs to me that the same thing might be true for three **non-parallel lines**. In other words, given three non-parallel lines, is there a unique circle which has the three lines as tangent?

Let's first have a look at the case where we have two parallel lines and one transversal (see Figure 2.7).

These three lines divide the plane into six regions. In region I, I can put lots of circles but none of them will have the **three** lines as tangents because only two lines define the region. The same is true for regions II, IV and V.

Do regions III and VI define a unique circle which has the three given lines as tangents?

Exercises

36. Show that regions III and VI each possess a unique circle with the three initial lines as tangents.
    Where are the centres of the two circles?
    What is the radius of each of the circles?
37. How many circles exist in regions I, II, IV and V, which have the two lines as tangents?
38. Consider three lines, no two of which are parallel.
    (a) Into how many regions do the three lines divide the plane?

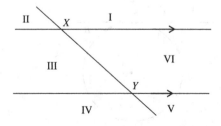

Figure 2.7.

(b) How many of these regions have a unique circle with the initial lines as tangents?

Where are the centres and radii of these circles?

(c) How many of these regions have an infinite number of circles with two of the initial lines as tangents?

Where are the centres of these circles?

In Figure 2.8 we show part of the situation where no two of the three lines are mutually parallel.

The three circles shown are external to $\triangle ABC$. They are therefore called the **excircles** of $\triangle ABC$. Their **excentres**, i.e. the centres of the excircles, are at I_a, I_b, I_c. We will denote the corresponding **exradii** by r_a, r_b, r_c.

We know from the way we constructed these excircles, that $I_a I_b$ is the perpendicular bisector of the external angle at C of $\triangle ABC$.

Before you get too worried about me calling that angle, **the** external angle at C, ask yourself what angles there can be that are external to $\triangle ABC$.

Consider Figure 2.9. I could have $\angle DCE$. But if I bisect this, I also bisect $\angle ACB$. So if, for some perverse reason, I want to bisect angles around C, bisecting $\angle DCE$ is no improvement on bisecting the internal angle $\angle ACB$.

On the other hand, if I bisect $\angle BCD$ I get the same line as when we bisect $\angle ACE$. This is a new line for me. So to all intents and purposes

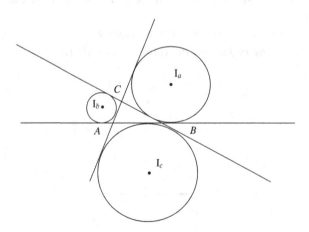

Figure 2.8.

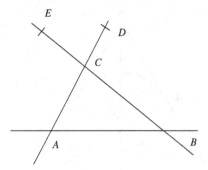

Figure 2.9.

I might as well think of $\angle ACE$ and $\angle BCD$ as the only worthwhile **external** angles of the $\triangle ABC$ at C.

Continuing in this vein, I can see that I_bI_c bisects the external angle at A and I_cI_a bisects the external angle at B.

Exercises

39. Let AB, AC meet the excircle with centre I_a at X, Y, respectively. Show that $AX + AY = a + b + c$ (where a, b, c represent the usual sidelengths in $\triangle ABC$).

40. Show that $\frac{1}{r} = \frac{1}{r_a} + \frac{1}{r_b} + \frac{1}{r_c}$, where r is the radius of the incircle of $\triangle ABC$ and the other "rs" are the radii of the three excircles.

41. Draw the triangle with vertices I_a, I_b, I_c.
    (a) Show that A, B, C are the feet of the altitudes drawn from I_a, I_b, I_c, respectively to the opposite sides of $\triangle I_aI_bI_c$.
    (b) Show that the orthocentre (see Exercise 32) of $\triangle I_aI_bI_c$ is the incentre of $\triangle ABC$.

42. If $\triangle ABC$ is an obtuse angled triangle, do all the results of the last exercise hold?

2.4. The 6-Point Circle?

In this section I want to explore another circle related to a triangle. I'll start this by looking at what I can get from an equilateral triangle. There is a diagram in Figure 2.10 that will help me focus my attention. I'll investigate the circle with centre O, that passes through the feet of the medians, L, M and N, on BC, CA, and AB, respectively.

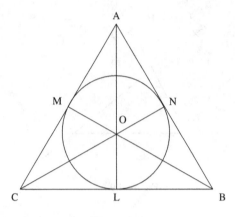

Figure 2.10.

Suppose I let the equilateral triangle have side length 2. In the diagram, AL, BM and CN are the medians of the triangle, so $AN = NB = \cdots = MA = 1$.

The first question that I want to ask is whether or not AL is perpendicular to BC. If it is, then OL is the radius of the circle we are looking at. Further we know from Exercise 31 that $\frac{AO}{OL} = 2$. So we might be able to find the radius of the circle in Figure 2.10.

Exercises

43. Show that, for an equilateral triangle, the medians are also altitudes.
44. Find the radius of the circle through L, M and N.
45. The circle and the line AO in Figure 2.10 intersect at a point. Call this point X. What is the length AX?
    Note that there are similar points Y and Z on BO and CO, respectively. What are the lengths BY and CY?

Having successfully noted that there is a circle related to equilateral triangles in a "natural" way, I want to switch my attention to right angled triangles. I'll use the standard $3, 4, 5$ triangle as a starting point but I'll scale it up to a $12, 16, 20$ triangle to avoid some fractions that would appear otherwise. Again, I know that any non-linear three points define a circle, so I'll continue to investigate the circle that is generated by the midpoints of the sides of our triangle. I've made a start in Figure 2.11, where the medians are drawn along with the circle through the midpoints L, M, N.

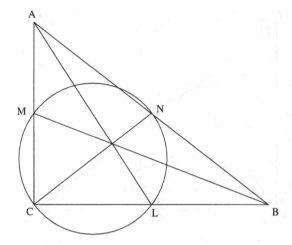

Figure 2.11.

First I want to find the centre and radius of this circle, and then I'll try to find other points on the circle. But this time, it doesn't look as if the centroid of the triangle (the point of intersection of the medians AL, BM and CN) is the centre of the circle.

Exercises

46. Where is the centre, O, of the circle through L, M and N in Figure 2.11? What is the radius of that circle?
47. What other points, that have some relevance to the triangle or to the lines we have drawn so far, lie on our circle?
48. Is there anything special about the 12, 16, 20 triangle? Can we repeat this investigation for **any** right angled triangle?

In Figure 2.12, we have added the line CS to Figure 2.11. Here I've called the point, other than N, where the circle and AB intersect, S. Is there anything special about this line? In the equilateral case that we considered earlier in this section, L, M and N were not only the midpoints of the respective sides, they were also the feet of the altitudes from the vertex to the opposite side. Could S be the foot of the altitude from C to AB?

Let's go the other way. Suppose that T is the foot of the altitude from C to AB. If we can show that T is on the circle then T will have to be S. But how can we do this? Look at Figure 2.13.

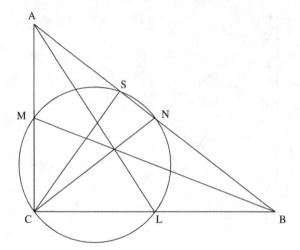

Figure 2.12.

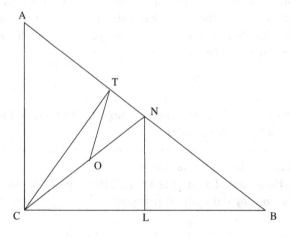

Figure 2.13.

In $\triangle CTO$, we're about to find OT by using the Cosine Rule. But this means that we have to find the angle at C as well as CO and CT.

Let's do some angle chasing. Let $\angle BAC = \alpha$ and let $\angle ABC = \beta$. Then since $\triangle ACT$ has a right angle at T and the angle α at A, then $\angle ACT = \beta$. Now it is easily seen that $\triangle CNL$ has sides $6, 8, 10$, so $\angle BCN = \beta$. This means that $\angle TCO = 90° - 2\beta$.

As for side lengths, $CO = 5$ because it is the radius of the circle. Now Δ's ABC and ACT are similar because of their angles. So $\frac{CT}{BC} = \frac{AC}{AB}$. Hence $CT = (16 \times 12)/20 = 48/5$.

Applying the Cosine Rule in ΔCTO we find that $OT = 5$ and T, the foot of the altitude from C to AB is on the circle. So $S = T$!

Exercises

49. Check the details of the Cosine Rule to make sure that $OT = 5$.
50. Does what we have done above apply to **every** right angled triangle ABC? Does the circle through the midpoints of the sides of **any** right angled triangle pass through the vertex, C, with the right angle and through the foot of the perpendicular from C to AB?
51. Is there anything interesting about U and V, the internal points where AL and BM, respectively, cut the circle?

Finally in this section, let's look at the isosceles triangle that has side lengths 2, 2 and $2\sqrt{3}$, see Figure 2.14. When this has been done, I'll look over what has been achieved by investigating these three examples.

Because of the symmetry of isosceles triangles, AL is both the median and the altitude from A to BC. In this triangle, the altitudes CT and BS meet AL outside the triangle at the orthocentre G. Where is the centre of the circle through L, M and N, and do S and T lie on this circle?

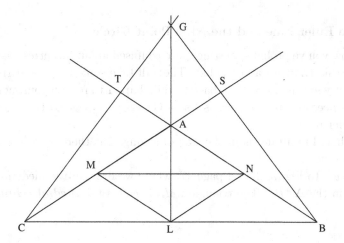

Figure 2.14.

Exercises

52. Where is the centre, O, of the circle through L, M and N and do S and T lie on this circle?
53. Are there any other interesting points on our circle?
54. What can be said about the centroids, the orthocentres and the circumcentre of the circles that we have considered in this section?

So what have we discovered in the examples we have looked at in this section? First we have built a circle that goes through all the midpoints of the sides. From that we have found that

(i) The feet of the altitudes from the vertices of the original triangle to the opposite sides are always on that circle.

But for the isosceles triangle, and you should check it for the other triangles, we have also found that

(ii) If we join the vertices of the triangle to the orthocentre of the triangle, that line meets the circle at points halfway between the vertex and the orthocentre.

Although we started looking at six points — the midpoints of the sides and the feet of the altitudes — we may have found nine distinguishable points that are actually on that circle.

It's also worth noting too, ... No, I'll get on to that in the next section.

2.5. The Euler Line and the Nine Point Circle

About now you've probably getting quite confused about incentres, circumcircles, orthic triangles and the like. They all take some time to learn and absorb but you can always look back to this chapter to refresh your memory when you need to. In fact there is also a Glossary on page 84 that you can always refer to.

All that, I'm afraid, is in the way of saying that there are some more names coming.

Go back to Figure 2.5 on page 47, where we drew in the medians AL, BM, CN in the $\triangle ABC$. The triangle LMN is called the ***medial triangle***.

Exercises

55. Prove that the centroid of $\triangle ABC$ is also the centroid of the medial triangle *LMN*.

56. Show that the orthocentre of the medial triangle is the circumcentre of the original triangle.

57. Show that the orthocentre, the centroid and the circumcentre of any triangle are collinear.

   Further show that the centroid divides the distance from the orthocentre to the circumcentre in the ratio 2:1.

The line joining the orthocentre, the centroid and the circumcentre is called the **Euler line** of the triangle.

Exercises

58. Show that the perpendicular bisector of *LM* in Figure 2.5 meets the Euler line halfway between the orthocentre and the circumcentre of △*ABC*.

59. Show that the midpoint of the Euler line segment between the orthocentre and the circumcentre, is the circumcentre of △*LMN*.

60. Show that the circumradius of the medial triangle is half the circumradius of the original triangle.

So now we know four points on the Euler line. These are the orthocentre, the centroid and the circumcentre of the original triangles as well as the circumcentre of the medial triangle.

Exercises

61. Where is the circumcentre of the medial triangle?

62. Find other points on the Euler line.

I now want to have a closer look at the circumcircle of the medial triangle. In Figure 2.15. △LMN is the medial triangle, H is the orthocentre, T is the centroid and O the circumcentre of △ABC.

I know from Exercise 59, that $PQ = QR$. Now PC', QQ' and RN are parallel and QQ' is halfway between the other two.

Now look at △'s $TC'Q'$, TNQ'. From above $\angle TQ'C' = \angle TQ'N = 90°$ and $C'Q' = Q'N$. Clearly TQ' is common to both triangles. Hence they are congruent (SAS). This means that $TC' = TN$.

Exercises

63. Find two other points which lie on the circumcircle of the medial triangle.

64. (a) In Figure 2.15, let Y be the point of intersection of NT and CH. Show that Y is on the circumcircle of the medial triangle.

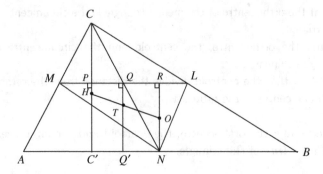

Figure 2.15.

(b) Let Z be the point of intersection of LT and AH. Show that Z is on the circumcircle of the medial triangle.

(c) Find a ninth point of the circumcircle of the medial triangle.

(d) Show that $CY = YH$.

As a result of the above discussion we now know nine points which lie on the circumcircle of the medial triangle. We will refer to this as the Nine Point Circle.

Nine Point Circle Theorem. *The three midpoints of the three sides of any triangle, the feet of the three altitudes of that triangle and the midpoints of the segments from the three vertices of the triangle to its orthocentre, all lie on a single circle.*

The centre of this circle lies on the Euler line, midway between the triangle's orthocentre and its circumcentre.

The radius of this circle is half of the circumradius of the triangle.

Exercise

65. In Figure 2.8 identify the nine point circle of $\triangle l_a l_b l_c$. Prove your assertion.

2.6. Some More Examples

In this section we give some old IMO problems or problems submitted for an IMO, that are related to the triangles and circles we've covered in this chapter.

Exercises

66. ABC is a triangle, the bisector of angle A meets the circumcircle of $\triangle ABC$ in A_1. Points B_1 and C_1 are defined similarly. Let AA_1 meet the line that bisects the two external angles at B and C, at A_0. Points B_0 and C_0 are defined similarly.

    Prove that area of $\triangle A_0 B_0 C_0 = 2 \times$ area of hexagon $AC_1 BA_1 CB_1$
    $$\geq 4 \times \text{ area } \triangle ABC.$$
    (IMO, 1989; submitted by Australia)

67. Triangle ABC is right angled at A and D is the foot of the altitude from A. The straight line joining the incentres of the triangles ABD, ACD intersects the sides AB, AC at the points K, L, respectively. S and T denote the area of the triangles ABC and AKL, respectively. Show that $S \geq 2T$.
    (IMO, 1988; submitted by Greece)

68. The triangle ABC is inscribed in a circle. The interior bisectors of the angles A, B and C meet the circle again at A', B' and C', respectively. Prove that the area of triangle $A'B'C'$ is greater than or equal to the area of triangle ABC.

    (Submitted by Canada in 1988)

69. Let ABC be an acute angled triangle. Three lines L_A, L_B and L_C are constructed through the vertices, A, B and C, respectively, according to the following prescription. Let H be the foot of the altitude drawn from the vertex A to the side BC; let S_A be the circle with diameter AH; let S_A meet the sides AB and AC at M and N, respectively, where M and N are distinct from A; then L_A is the line through A perpendicular to MN. The lines L_B and L_C are constructed similarly. Prove that L_A, L_B and L_C are concurrent.

    (Submitted by Iceland in 1988)

70. The triangle ABC has a right angle at C. The point P is located on segment AC such that triangles PBA and PBC have congruent inscribed circles. Express the length $x = PC$ in terms of $a = BC, b = CA$ and $c = AB$.
    (Submitted by USA in 1988)

71. Vertex A of the acute triangle ABC is equidistant from the circumcentre O and the orthocentre H. Determine all possible values for the measure of angle A.

    (Submitted by USA in 1989)

2.7. Hints

1. Draw the diagram of each situation. You know from Chapter 5 of *First Step* that the circumcentre is on the perpendicular bisector of each of the three sides.

   (a) At the midpoint of the hypotenuse. Why?

   (b) How far is it from each vertex?

   (c) All we can really say is that it is somewhere on the axis of symmetry.

2. Is there only one possibility?

   Draw a diagram showing where the perpendicular bisectors go.

3. If you can't see the answer straight away, see what happens in isosceles triangles. Let the angle which is different from the other two be θ. We know how big θ is when the circumcentre is on one of the sides. How big is it when circumcentre is inside? What about outside?

   Does this generalize to **any** triangle?

4. (a) That should be easy given Exercise 1(a).

   (b) Exercise 1(b) should give this away.

5. Does it look as if $\angle AB'C = \angle ABC$? Can $\angle AB'C = 90°$?

6. Do they seem to be the same again? Try several different positions for C'. Do you always get $\angle AC'B = \angle ACB$?

7. Will $\angle AC'B = \angle ACB$, no matter what $\angle ACB$ is? Try some more examples.

8. Haven't we done this somewhere before?

9. You now have to show that if $AB = A'B'$, then $\angle ACB = \angle A'C'B'$. Draw diameters BD and $B'D'$.

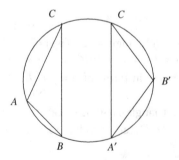

10. Well, no. Clearly D is on the wrong side of AB. Try a few examples to see what $\angle ADB + \angle ACB$ equals. Surely it's not ...

11. Draw a diagram and argue from there.

12. Equality didn't hold in Exercise 10, so why should it hold now? However, Exercise 10 suggests that $\angle ABC + \angle ADC = 180°$. Does experimental evidence suggest this is a worthwhile conjecture? Can it be proved?

13. What other perpendiculars are associated with the circumcircle?

14. Answer the questions in the last bit of text. By the way, how big is AA'?

15. See the solution of Exercise 14.

16. See the solution of Exercise 14.

17. $\Delta = \frac{1}{2}$ base $\times$ altitude $= \frac{1}{2} c \times h$, where h is the perpendicular from C to AB.

18. Draw the obtuse angled triangle case. The right angle case ought to be easy.

19. (a) Use the fact that if $\angle XAC = 90°$, then AC is a diameter.

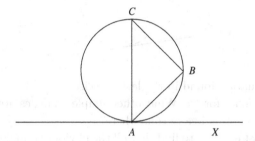

(b) Should follow easily from (a) but you might need to put another line in the diagram.

20. The diagram is shown below. Construct the point B' such that BB' is a diameter. $c = AB = 2q \sin \angle AB'B$. What is $\angle AB'B$?

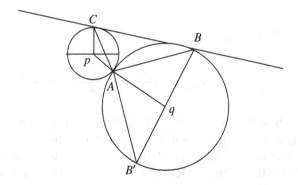

21. Here is the diagram. Now use congruence.

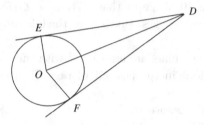

22. You could look back to Chapter 5 of *First Step*.
23. Add up the area of Δ's *AIB, BIC, CIA*.

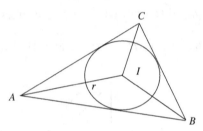

24. Apply the equation found in the last exercise.
25. Think right angles for $\frac{5}{2}r$. What other simple triangles are there?
26. Haven't we done this before somewhere?
27. Repeat the last exercise to find that all the obvious triangles are similar.
28. The quick proof is via parallelograms.
29. Similarity!
30. We know that $x = y$ and $v = w$. Is $v = x$?
31. For $\frac{AG}{GL}$ use Δ's *AGB, LGB*.
32.

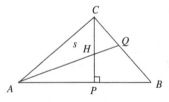 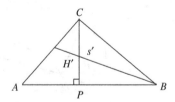

  If you can show that $s = s'$, then the points H, H' are identical. This will then get the result we want. Use some trigonometry.

33. Complete one of the diagrams of the last exercise to show the orthic triangle *PQR*. If ∠*QPH* = ∠RPH, then the incentre of △*ABC* lies on the altitude *CP*.
  (Is *APHR* a cyclic quadrilateral?)

34. Compare angles.
35. (a) Not much of a triangle.
    (b) Yes.
36. The tangents will meet at X and Y. So it's angle bisecting time again.
37. How many would you like?
38. (a) Take a triangle and extend the sides to infinity.
    (b) How many unique centres can you find?
    (c) This must be the rest. But where could the centres be?
39. Let Z be the point where the excircle with centre I_a touches the tangent BC.
40. First get a relation between Δ and r_a similar to that for Δ and r, viz., $\Delta = \frac{1}{2}(a + b + c)r$. Then repeat for Δ and r_b, and Δ and r_c.

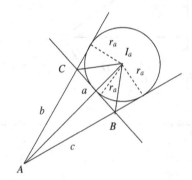

41. It's all about bisecting angles again.
42. Why not? Draw a diagram and see.
43. Congruent triangles?
44. What do you know about the centroid or maybe just side lengths.
45. Use known side lengths.
46. What can you say about the quadrilateral $CLNM$?
47. What things are special about a triangle?
48. Repeat the arguments of the 12, 16, 20 triangle and see what happens.
49. State the Cosine Rule and substitute what you know here.
50. Go back to Exercise 45.
51. Do they represent anything that might be generalized?
52. Perpendicular bisectors. But of what?
53. Sure are!
54. Ah now, what might we expect? They are on a circle? Put the points in on a diagram and see what it looks like.
55. Just recall what all the definitions are.

56. Definitions again.

57. This is a little tricky. The basic aim is to show that two triangles are similar.

    First construct the diagram.

    To show: O, G, H are collinear.

    This can be done if we can show that $\triangle$'s CHG, NOG are similar and the sides are in the ratio 2:1.

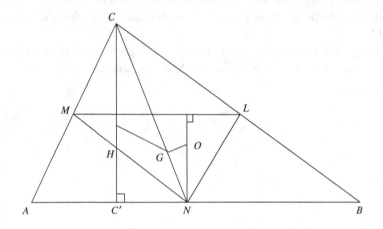

    Why is $CG : GN = 2 : 1$?

    Why is $CH : NO = 2 : 1$? (Orthocentres and ratios.)

    Why is $CH \,\|\, ON$?

58. Take part of the diagram of the last exercise. Are $\triangle$'s CWX, NVX similar or even congruent?

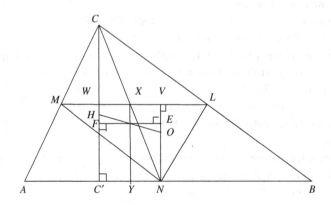

Let XY be the perpendicular to LM (and AB) and let XY intersect the Euler line at T. Are Δ's OET, HFT congruent?

59. Similar to the last exercise.

60. Similarity on the old scale of 2:1.

61. Let $\Delta L'M'N'$ be the medial triangle of ΔLMN. The circumcentre of $\Delta L'M'N'$ is halfway between the orthocentre and the circumcentre of ΔLMN. What is this to ΔABC?

62. In general, there are zillions of such points.

63. Rotate Figure 2.15.

64. (a) Look for similar triangles again.

    (d) Consider Δ's CMY, CAH.

65. What is the medial triangle for $\Delta I_a I_b I_c$?

66. How are the circumcircle of Δ's ABC and $A_0B_0C_0$ related? If I is the intersection of AA_0 and BB_0, how are the areas of Δ's IA_1B and A_0A_1B related?

67. Show that the line joining the centres of the two circles and AC form an angle of $45°$.

68. Use the hexagon idea from Exercise 66.

69. Focus your attention on the circumcentre of Δ ABC.

70. What do you know about size of the inradius of a triangle? Find an expression for x. From here it is almost pure algebra.

71. Use trigonometry to find the unique value of A.

2.8. Solutions

1. (a) Let P be the midpoint of BC and draw PQ perpendicular to BC to meet AC at Q. Then draw QR perpendicular to AB. Δ's PQC, RAC are similar (why?) and since $BP = RQ$ (why?), they are in fact congruent. Hence Q is the midpoint of AC.

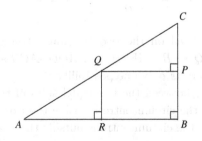

The perpendicular bisectors of the sides must therefore meet at Q and you have found your circumcentre.

(b) The perpendicular bisectors of the sides of an equilateral triangle bisect the opposite angles. (Why?) Let the sidelength of the triangle be ℓ. Then $CR = \frac{1}{2}\sqrt{3}\ell$ and $AR = \frac{1}{2}\ell$.

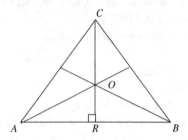

In $\triangle ARO, \angle RAO = 30°$. Hence $\frac{OR}{AR} = \frac{1}{\sqrt{3}}$. So $OR = \frac{1}{2\sqrt{3}}\ell$. Thus $CO = \frac{\sqrt{3}}{3}\ell$.

We have thus located the circumcentre O and also found the circumradius. By symmetry, $AO = BO = CO$. (What is the circumradius of a right angled triangle?)

(c) If AC = CB, then the perpendicular bisector of AB goes through C. The centre must be on this bisector. We can say no more than this without knowing the angles in $\triangle ABC$. However, we can say when is the centre inside $\triangle ABC$, when is it outside $\triangle ABC$, and when is it on an edge of $\triangle ABC$.

2. Suppose O is the circumcentre on side AB. Let Q be the midpoint of AC. Since O is the circumcentre, QO is perpendicular to AC.

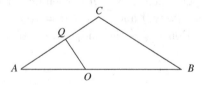

The Cosine Rule used on the two "obvious" triangles, can be used to show that $2OQ = BC$. Hence $\triangle$'s AQO, ACB are similar. Hence $\angle ACB = 90°$. There *is* only one possibility.

3. Let θ be the angle between the two equal sides of the triangle. If θ is smaller than $90°$, the circumcentre is inside the isosceles triangle. If θ is larger than $90°$, the circumcentre is outside the triangle.

Does this mean that in an acute angled triangle (where all angles are smaller than 90°), the circumcentre is inside the triangle? So for obtuse angled triangles, is the circumcircle **outside** the triangle?

I settle this question later on. Can you settle it now? If necessary, revisit this problem after completing Exercise 8.

4. (a) Half the length of the hypotenuse.
   (e) We know that the circumradius is $\frac{\sqrt{3}}{3}\ell$, where ℓ is the length of the side of the equilateral triangle.

5. What is the circumcentre of $\triangle AC'B$? So it's likely that $\angle AC'B = 90°$.

6. If you're measuring the angle, then $\angle AC'B$ probably won't always be 60°. However it ought to be close enough to conjecture that $\angle AC'B = \angle ACB = 60°$. So how can this be proved?

   Complete the diagram below, where $\alpha = \angle AOB, \beta = \angle DOB, \gamma = \angle OC'A, \delta = \angle OAC'$ and $\varepsilon = \angle OC'B$. Now $\angle OAB = \angle OBA = 30°$, since we know that the circumradii OA and OB, bisect $\angle$'s CAB, CBA, respectively. But $\gamma = \delta$ since $\triangle AOC'$ is isosceles. Further $\alpha = \gamma + \delta$ (external angle equals the sum of the two interior and opposite angles).

   Hence $\gamma = \delta = \frac{1}{2}\alpha$.

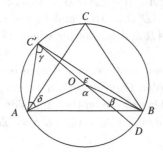

Similarly $\varepsilon = \frac{1}{2}\beta$. So $\angle AC'B = \gamma + \varepsilon = \frac{1}{2}(\alpha + \beta)$.

Finally in $\triangle AOB, \angle AOB = 180° - \angle OAB - \angle OBA = 120°$. Hence $\angle AC'B = \frac{1}{2}(\alpha + \beta) = \frac{1}{2}120° = 60°$.

(For another solution to this exercise, see Exercise 6 revisited on page 45.)

7. You probably know $\angle AC'B$ always equals $\angle ACB$. But can you prove it? If you can, you can check out your answer in the next piece of text.

8. This is proved in part of Exercise 6, where $\alpha + \beta$ happens to be 120°. However, the fact that $\gamma + \varepsilon = \frac{1}{2}(\alpha + \beta)$ doesn't rely on the fact that $\alpha + \beta = 120°$ but only on $\alpha + \beta < 180°$. Now show the same proof holds for $\alpha + \beta \geq 180°$.

9. The first part of this follows directly from the last exercise.

Now let AB and $A'B'$ be equal chords on a circle and let D and D' be such that BD and $B'D'$ are diameters. Δ's ABD, $A'D'B'$ are congruent (RHS — right angle, hypotenuse, side). So $\angle ADB = \angle A'D'B'$. From the text and Exercise 8, we know that, if C and C' are subtended by AB and A'B', respectively, that $\angle ADB = \angle ACB$ and $\angle A'D'B = \angle A'C'B'$. Hence $\angle ACB = \angle A'C'B'$ as required. (And this is always true since the angle subtended on the circle is always twice the angle subtended at the centre.)

It doesn't matter where that fixed chord is since the triangles formed by the endpoints of the chord and the centre of the circle are congruent. Then use "the angle subtended at the centre" property.

Hmm. But should you be worried about where all of this subtending is going on? Check out Theorem 2.

10. The two angles add to 180°. One way to see this is to draw the perpendicular bisector of AB. Let this intersect the circle at $C'D'$. Hence $C'D'$ is a diameter of the circle, so the angle at A in $\Delta AC'D'$ is 90°. That means that the angles at $C'D'$ add to 90°. Repeating this with $\Delta BC'D'$ gives $\angle AC'B + \angle AD'B = 180°$. So by the last exercise we have $\angle AC'B + \angle AD'B = 180°$ and this is true no matter where C is on one side of AB and no matter where D is on the other side of AB.

This discussion means that we should have been a bit more careful about the way we worded Exercise 8 than we were. It's important that the angles we are looking at are on the same side of the chord.

11. Since the circumcentre of ΔABC is on AB, then ΔABC is right angled at C.

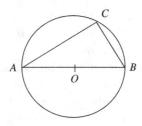

Alternatively, since $\angle ACB = \frac{1}{2}\angle AOB$ (the angle subtended at the circle is half that subtended at the centre) and since $\angle AOB = 180°$, $\angle ACB = 90°$.

12. Now $\alpha + \beta = 360°$. Further $\angle ABC = \frac{1}{2}\alpha$ and $\angle ADC = \frac{1}{2}\beta$. Hence $\angle ABC + \angle ADC = 180°$. The same argument holds for $\angle BAD + \angle BAC$. So the sum of opposite angles of a cyclic quadrilateral is $180°$.

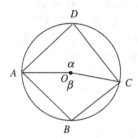

Alternatively this is a direct result of Exercise 10 using AC as a fixed chord.

13. This is based on the fact that the perpendicular bisector of AB goes through the circumcentre. Since OL goes through O and is perpendicular to AB it must be the perpendicular bisector of AB.

   You can also use a congruent triangle argument.

14. In Figure 2.3, $\triangle ABA'$ is right angled at B and $\angle AA'B = \angle ACB$. So $\frac{AB}{AA'} = \sin C$. However, $AA' = 2R$ and $AB = c$. Hence $\frac{c}{\sin C} = 2R$.

15. 2R again. The proof is analogous to that of the last exercise. This time, first draw the triangle $CC'A$, where C, C' are the opposite ends of a diameter of the circumcircle.

16. This comes after drawing the diameter BB'. From there, the proof is analogous to the last two Exercises.

17. Let h be the altitude from C to AB. Then $\Delta = \frac{1}{2}AB \cdot h$. But $AB = c$ and $h = b\sin A$, where $b = AC$. Since $\sin A = \frac{a}{2R}$ we get $\Delta = \frac{1}{2}cb(\frac{a}{2R})$. Rearranging gives $R = \frac{abc}{4\Delta}$. (Naturally any side of the triangle could have been taken as the base.)

   What is R for the specific cases of a right angled triangle and an equilateral triangle?

18.

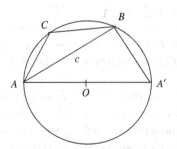

From $\triangle AA'B$, $\frac{c}{2R} = \sin A' = \sin(180° - C)$ by Exercise 12. However, $\sin(180° - C) = \sin C$. So the result $\frac{c}{\sin C} = 2R$ holds here too. Now check out $\frac{a}{\sin A}$ and $\frac{b}{\sin B}$. (This is done later.)

19. (a) $\triangle ABC$ has a right angle at B since CA is the diameter of the circumcircle. Hence $\angle XAB + \angle BAC = 90° = \angle ACB + \angle BAC$ (using $\triangle ABC$). Hence $\angle XAB = \angle ACB$.

   (b) In the general case, construct the diameter AA' of the circumcircle through A. From (a), $\angle XAB = \angle AA'B$. But $\angle AA'B = \angle ACB$ since they are the same chord.

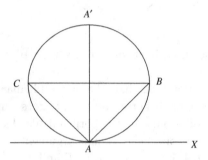

20. The diagram is shown below.

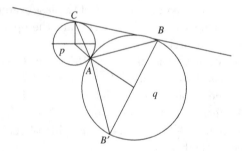

By Exercise 19, $\angle AB'B = \angle ABC = B$. From Exercise 18 and $\triangle ABB'$, $c = 2q \sin B$. Similarly $b = 2p \sin C$.

   Hence $pq = \frac{1}{4}(\frac{b}{\sin B})(\frac{c}{\sin C}) = \frac{1}{4} \cdot 2R \cdot 2R = R^2$.

21. $\triangle$'s DEO, DFO are congruent. This is because $\angle DEO = \angle DFO = 90°$, $DO = DO$ and $EO = OF$ because they are radii of the circle. Hence $\triangle$'s DEO, DFO are congruent by the RHS test (see Chapter 5 of *First Step*).

   From the above we see that $DE = DF$ and $\angle ODE = \angle ODF$.

22. From the last exercise you should be able to see that the centre of the incircle lies on the bisectors of each of the angles of the triangle. So bisect the angles of the triangle (the method is shown in Chapter 5

of *First Step*). Where these bisectors meet is the incentre, I. Drop the perpendicular from I to any side of the triangle to obtain the inradius.

23. The area of $\triangle AIB = \frac{1}{2}cr$. Similarly, area $\triangle BIC = \frac{1}{2}ar$ and area $\triangle CIA = \frac{1}{2}br$. Hence $\triangle = \frac{1}{2}r(a+b+c)$, where $\triangle$ is the area of $\triangle ABC$.

24. $\triangle = 6, a = 3, b = 4, c = 5$, so $r = 1$. (Is there a way of doing this directly?)

25. $R = \frac{5}{2}r$ for the 3, 4, 5 triangle. After all, we showed that $r = 1$ and R is just half the length of the hypotenuse.

    For equilateral triangles, $R = 2r$. (This can be checked out directly from the formulae for R and r which involve $\triangle$.)

    For what right angled triangles is $R = 2r$?

26. $\triangle$'s CML, CAB are similar because they both have the angle C and corresponding sides are in the same ratio $(1 : 2)$. Hence $\angle CML = \angle CAB$. So $ML \| AB$.

27. Using the arguments of the last Exercise, $\triangle$'s MNA, LNB are congruent to $\triangle CML$ and similar to $\triangle CAB$. $\triangle NLM$ is congruent to $\triangle CML$ by the SSS argument. Hence we have four congruent triangles in $\triangle ABC$.

28. Since the diagonals of a parallelogram bisect each other, AL must bisect MN ($AMLN$ is a parallelogram since $ML \| AN$ and $ML = AN$).

29. $\triangle$'s CXL, CYB are similar because they both have the angle C and $\angle CLX = \angle CBY$. But $CL : CB = 1 : 2$, so $CX : CY = 1 : 2$.

30. $v + w = 2x$ and $v = w$. Hence $v = x$. So $v = w = x = y$ and all the triangles have the same area.

31. $\triangle$'s AGB, LGB have the same altitude. Any difference in their areas will be caused by the difference in their bases, AG and LG. Now $\frac{\text{area } \triangle AGB}{\text{area } \triangle LGB} = \frac{2x}{y} = 2$, since $x = y$. Hence $\frac{AG}{LG} = 2$.

    Similarly, it can be proved that $\frac{BG}{GM} = \frac{CG}{GN} = 2$. Hence the centroid of a triangle is two thirds of the way along a median (from a vertex).

32. Let $\alpha = 90° - A$, $\beta = 90° - B$, $\gamma = 90° - C$, $s = CH$, and $s' = CH'$. In $\triangle AQC, \angle CAQ = 90° - C = \gamma$ and $CQ = AC \sin\gamma$. From $\triangle CPB$, $\angle BCP = 90° - B = \beta$.

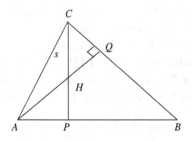
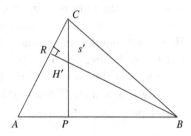

Using $\triangle CHQ$, this gives $s = CQ \sec \beta$. Hence $s = AC \frac{\sin \gamma}{\cos \beta} = b \frac{\cos C}{\sin B}$.
Similarly, from $\triangle CH'R$ we get $\frac{CH'}{\cos C} = \frac{a}{\sin A}$. By the Sine Rule we then
see that $s = \frac{b \cos C}{\sin B} = \frac{a \cos C}{\sin A} = s'$.

34. Let P, Q, R be the feet of the altitudes from C, A, and B, respectively. It
can be shown that $\angle CAQ = \gamma = \angle CBR$. Now consider the circumcircle
of $\triangle AHR$. This has AH as diameter and so it must be the circumcircle
of $\triangle AHP$ as well. On this circle, $\angle HAR$ and $\angle HPR$ subtend the same
arc. Hence $HPR = \gamma$.

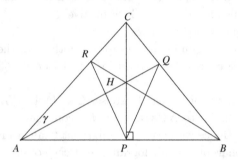

Using an analogous argument on the quadrilateral $BQHP$, we find
that $\angle HPQ = \gamma$ too. Hence the incentre of the orthic triangle lies on
the altitude CP.

Similarly we can show that $\angle RQH = \angle PQH$ and $\angle QRH = \angle PRH$.
So the incentre of the orthic triangle lies on the other two altitudes (AQ
and BP). Hence the incentre of $\triangle PQR$ lies on the intersection of the
altitudes of $\triangle ABC$. In other words, the incentre of the orthic triangle
is the orthocentre of the original triangle.

34. From the last exercise we can see that angles A, B, C in the original
triangle go to angles $180° - 2A, 180° - 2B, 180° - 2C$ in the orthic
triangle. If $180° - 2A = A$, then $A = 60°$. Hence $180° - 2B = B$ and
$180° - 2C = C$ or $180° - 2B = C$ and $180° - 2C = B$. In both cases
$B = C = 60°$.

Essentially the only remaining case is $180° - 2A = B, 180° - 2B =$
$C, 180° - 2C = A$. Algebra again gives $A = B = C = 60°$.

Hence only equilateral triangles have similar orthic triangles.

35. (a) If the right angle is at A, the orthic triangle is AAP, where P is
the foot of the altitude from A to BC. This means that the orthic
triangle is a line! (Does anything like this happen for any other
triangle?)

(b) Suppose $A = 45°$. Then $180° - 2A = 90°$. So if one of the angles of the original triangle is $45°$, then the orthic triangle is right angled. UNLESS?

36. If a circle exists in region III, two of its tangents pass through X. Hence the centre of any such circle is on the bisector of the angle at X. But the centre is also on the bisector of the angle at Y. Since these two bisectors cannot be parallel, they must meet at a unique point. Hence if there is more than one circle, they must all share the same centre.

   Now a diameter of a circle exists which is perpendicular to the two original parallel lines. Hence the diameter of a circle is equal to the perpendicular distance between the two parallel lines. So there is only one circle. Its centre is on the bisectors of X and Y and its radius is a half the distance between the two original parallel lines.

   The same arguments apply to region VI.

   Under what conditions will the circles produced for regions III and VI touch?

37. Once again the centres of these circles are on the bisectors of the appropriate angles at X or Y (but not both). There are no other restrictions, so there are an infinite collection of circles in each region.

38. (a) Seven.

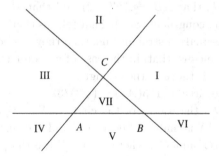

(b) Unique circles exist in I, III, V and VII. The argument for unique centres is essentially the same as in Exercise 36.

   I think that the radius for the circle in I is $\dfrac{AC \tan \frac{C}{2}}{(1-\tan \frac{A}{2} \cot \frac{C}{2})}$.

   In the region VII we have the incircle which is unique. We have already found the inradius in terms of the area and the lengths of the sides.

(c) Regions II, IV, VI have an infinite collection of circles. The centres are again on the appropriate angle bisectors.

39. By previous arguments regarding tangents from a point, $BX = BZ$ and $CY = CZ$. Then $AX + AY = AB + BX + AC + CY = AB + AC + (BZ + CZ) = a + b + c$.

40. Area $\triangle ABI_a = \frac{1}{2}cr_a$. Area $\triangle ACI_a = \frac{1}{2}br_a$. Area $\triangle BCI_a = \frac{1}{2}ar_a$. Hence $\Delta = \text{area}\,\triangle ABC = \text{area}\,\triangle ABI_a + \text{area}\,\triangle ACIa - \text{area}\,\triangle BCI_a = \frac{1}{2}(b+c-a)r_c$. So $\frac{1}{r_a} + \frac{1}{r_b} + \frac{1}{r_c} = \frac{1}{2\Delta}\{(b+c-a\} + (a+c-b) + (a+b-c)\} = \frac{a+b+c}{2\Delta} = \frac{1}{r}$.

41. (a) Now I_bI_c bisects the external angles at A and I_aA bisects the internal angle at A (since AB, AC produced are tangents to the excircle with centre I_a).

    So $(\angle I_aAB + \angle BAI_c) + (\angle I_aAC + \angle CAI_b) = 180°$.

    Hence $\angle I_aAI_c = \angle I_aAB + \angle BAI_c = 90°$. But $\angle I_aAB = \angle CAI_a$ and $\angle BAIc = \angle CAI_b$, so AL is perpendicular to I_bI_c and so $\angle I_aAI_c = 90°$.

    Thus I_aA is perpendicular to I_bI_c. The fact that I_bB, I_cC are altitudes follows similarly.

    (b) I_aA, I_bB, I_cC intersect at the orthocentre of $\triangle I_aI_bI_c$ by (a). Since I_aA, I_bB, I_cC are also the bisectors of angles A, B, C, respectively, they also intersect at the incentre of $\triangle ABC$. The result thus follows.

42. Yes. There should be no change in any of the arguments.

43. Looking at $\triangle$'s ABL, ACL in Figure 2.10, we see that they have a common side, AL, that $AB = AC(= 2)$ and that $BL = CL(= 1)$. So $\triangle$'s ABL, ACL are congruent (SSS). That tells us that $\angle ALB = \angle ALC$. Since these angles make a straight line, then they are both right angles.

    Note that this means, that, in the case of an equilateral triangle, the medians are also altitudes of the triangle.

44. The radius of the circle is $1/3(AL) = (\sqrt{3})/3$.

45. Since $AX = AO-$ the radius of the circle, $AX = (\sqrt{3})/3$. Note that this means that AX is halfway between A and the centre O. A similar results hold for BY and CZ, where Y and Z are the points where BO and CO, respectively, intersect the circle.

46. Note that $MNLC$ is a rectangle because $MN \| CB$ and $NL \| AC$ (see Exercise 26). Since the angle at L is $90°$, ML is a diameter of the circle. Hence C is on the circle (see Exercise 11). So the centre of the circle is at the intersection of the two diagonals of that rectangle. So O is 4 horizontal units and 3 vertical units from C. So OC, the radius of the circle is 5 units. (Perhaps you now can see why we used 12, 16, 20 as the sides of the original right angled triangle.)

47. Clearly C lies on the circle because it is a vertex of the rectangle $MNLC$. But does C have any relevance apart from the fact that it is at the right angle of $\triangle ABC$?

   And what can we say about (i) the unnamed point where the circle and $\triangle ABC$ intersect, call this point S; (ii) the point (not L) where the circle and AL intersect, call this point U; and (iii) the point (not M) where the circle and BM intersect, call this point V?

48. See what you can do with this. We'll return to it later.

49. Now $OT^2 = OC^2 + CT^2 - 2 \times OC \times CT \cos(90° - 2\beta) = 5^2 + (48/5)^2 - 2 \times 5 \times 48/5 \times \cos(90° - 2\beta)$. So we had better worry about that cosine term. Now $\cos(90° - 2\beta) = \sin 2\beta = 2 \sin \beta \cos \beta$. And we can get these values form the original triangle: $\sin \beta = 12/20$ and $\cos \beta = 16/20$. This then gives us

$$OT^2 = 5^2 + (48/5)^2 - 2 \times 5 \times 48/5 \times 2 \times 12/20 \times 16/20$$

$$= 5^2 + (48/5)^2 - 2 \times 5 \times 48/5 \times 2 \times 3/5 \times 4/5$$

$$= 5^2 + (48/5)^2 - 2 \times 48/5 \times 2 \times 3 \times 4/5$$

$$= 5^2 + (48/5)^2 - (48/5)^2 = 5^2.$$

(Note the value of not simplifying the expression until the last moment.)

   Miraculously $OT = 5$ and so S is the foot of the altitude from C to AB. How about that?

(But there is a better way to do this problem.)

50. This follows on from the last Solution. It is just a matter of changing the lengths of the sides to variables a, b, c, say.

51. I can't find anything but if you do, please let me know.

52. O has to be on the intersection of the perpendicular bisectors of LM, MN and NL. AL is the perpendicular bisector of MN, so O has to be somewhere on AL.

   Let F be the midpoint of LN. Then $LF = FN = 1/2$. Since $\angle ALN = 60°$, $\triangle OFL$ has sides $1/2$, 1, and $\sqrt{3}/2$. So O is on AL and 1 unit up from BC so the radius of the circle is 1. But A is 1 from BC on AL, so $O = A$.

   Since $\triangle BCT$ is a right angled triangle with right angle at T, and $\angle TBC = \angle ABC = 30°$, $\angle BCT = 60°$. Hence $\angle ACT = 30°$ and $\angle CAT = 60°$. But $AC = 2$, so $AT = 1$. Hence T does lie on the circle centre O. By symmetry, so does S.

53. It's not too difficult now to show that $AG = 2$. So there is another point on the circle, R, where $TR = 1 (= RG)$.

   In this case it's worth noting that $CT = TG = \sqrt{3}, T$ and S, as well as R are half way between a vertex of the triangle and the orthocentre.

54. Do they lie on a line? See the next section for a proof for any triangle.

55. The centroid G, of $\triangle ABC$ is the point of intersection of AL, BM, CN. Since AL bisects MN, then part of AL is one of the medians of the medial triangle LMN. Similarly BM and CN are parts of the medians from M and N, respectively, to the medial triangle. Hence G is the centroid of the medial triangle.

56. Let LX be the perpendicular bisector of BC. Since $MN\|BC$, LX is also perpendicular to MN. Hence part of LX is the altitude from L to MN.

    The perpendicular bisectors of the sides of $\triangle ABC$ meet at the circumcentre of $\triangle ABC$. However, part of these bisectors are also altitudes of the medial triangle LMN. These altitudes meet at the orthocentre of $\triangle LMN$. Hence the orthocentre of the medial triangle is the circumcentre of the original triangle.

57. Now $CG:GN = 2:1$ is the old result of Exercise 32.

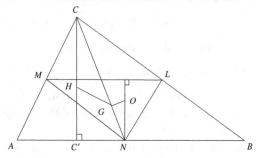

    Now let H be the orthocentre of $\triangle ABC$ and O be the orthocentre of $\triangle LMN$. But $\triangle$'s ABC, LMN are similar with $AB:LM = 2:1$. Hence $CH:NO = 2:1$ because H and O are the respective orthocentres of $\triangle$'s ABC, LMN.

    Since O is the orthocentre of $\triangle LMN$ and $AB\|LM$ (see Exercise 26), NO is perpendicular to AB. Hence $CH\|ON$. But this gives $\angle GCH = \angle GNO$, alternate angles on parallel lines.

    As a result of the above, $\triangle$'s CHG, NOG are similar. Hence $\angle OGN = \angle HGC$. Since CG is a straight line, the fact that $\angle OGN = \angle HGC$ implies (at last) that O, G, H are collinear.

    It's obvious from the ratio of $2:1$ in the similarity of the triangles, that $HG:GO = 2:1$. So the centroid divides the distance from the orthocentre to the circumcentre in the ratio of $2:1$.

58. Since $\triangle$'s ABC, LMN are congruent and $AB:LM = 2:1$, then the altitude from C to AB is twice the length of the altitude from N to LM. Hence $CW = NV$. $\triangle$'s CWX, NVX are therefore congruent and $WX = VX$.

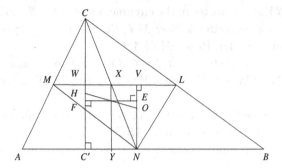

In △'s *OET*, *HFT*, we have right angles at *E* and *F* and ∠*OTE* = ∠*HTF* (vertically opposite). Further *FT* = *WX* = *VX* = *TE*. So △'s *OET*, *HFT* are similar. Hence *OT* = *TH*.

We have thus shown that the perpendicular bisector of *LM* meets the Euler line halfway between the orthocentre and the circumcentre of △*ABC*.

59. By an argument similar to that in the last exercise, the perpendicular bisector of *LN* (or of *MN*) meets the Euler line halfway between *O* and *H*. Hence all perpendicular bisectors of △*LMN* meet at *T*. This is therefore the circumcentre of the medial triangle.

60. The ratio of the sides of the triangle to the sides of the medial triangle is 2 : 1. Since $R = \frac{abc}{4\triangle}$, if we divide the sidelengths by 2 we divide the circumradius by 2. Hence the circumradius of the medial triangle is one half the circumradius of the original triangle.

61. The orthocentre of △*LMN* is the circumcentre of △*ABC* (see Exercise 56). This is *O* in the diagram of Exercise 58 (Hints).

   The circumcentre of △*LMN* is *T*. So the circumcentre of △*L'M'N'* is halfway between *O* and *T* on the Euler line of △*ABC*.

62. Just keep taking the circumcentre of the medial triangle of the medial triangle of the medial triangle of the...

63. The feet of the altitude from *A* and *B* to the opposite sides of △*ABC* must also lie on the circumcircle of the medial triangle.

64. (a) Let *T'* be on *CC'* such that *TT'* is perpendicular to *CC'*. Then △'s *YTT'*, *TNQ'* are congruent. Hence *YT* = *TN*. Since *TN* is a radius of the circumcircle of the medial triangle and *Y*, *T*, *N* are collinear, *YN* is a diameter of this circumcircle. Hence *Y* is on the circle too.

   (b) The same argument as for *Y* but looked at from the point of view of *A* and *L* rather than *C* and *N*.

   (c) The point of intersection of *MT* and *BH*.

(d) Since NY is a diameter of the circumcircle of $\triangle LMN$, $\angle YMN = 90°$
   (angle in a semi-circle). Now $MN \parallel BC$, so MY is perpendicular to
   BC. But so is AH. Hence $MY \parallel AH$.

   Now we can deduce that $\triangle$'s CMY, CAH are similar.

   So $CM : CA = 1 : 2$, then $CY : CH = 1 : 2$. So Y is the midpoint
   of CH.

65. $\triangle ABC$ is the medial triangle for $\triangle I_a I_b I_c$. Hence the nine point circle
   of $\triangle I_a I_b I_c$ is the circumcircle of $\triangle ABC$.

66. Let I denote the intersection of the three internal bisectors. Then I
   claim that

$$IA_1 = A_1 A_0. \tag{1}$$

One way of proving (1) is to realize that $A_0 A, B_0 B, C_0 C$ are altitudes
of triangle $A_0 B_0 C_0$, consequently the circumcircle of triangle ABC is
the nine point circle of triangle $A_0 B_0 C_0$ so it bisects the distance IA_0.
Or, without reference to the nine point circle:

$$\angle A_1 IB = \frac{1}{2}A + \frac{1}{2}B \text{ (external angles of } \triangle AIB)$$

$$\angle IBA_1 = \frac{1}{2}B + \frac{1}{2}A = \angle A_1 IB.$$

So $IA_1 = A_1 B. \tag{2}$

But also

$$\angle A_1 A_0 B = 90° - \angle A_1 IB,$$

$$\angle A_1 BA_0 = 90° - \angle IBA_1,$$

so $\qquad\qquad A_1 B = A_1 A_0. \tag{3}$

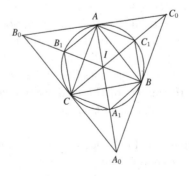

Considering (1) we get

$$\text{area } \Delta IA_1B = \text{area } \Delta A_0A_1B.$$

Repeating this argument for the six triangles that have a vertex at I and adding them together gives the required equality.

To prove the inequality, draw the three altitudes in triangle ABC, intersecting in H. Let X be the mirror image of H on the side BC, Y its mirror image on AC and Z on AB. Then X, Y, Z are points on the circumcircle of ΔABC. (This is because $\angle CXB = \angle CHB = 180° - A$ from the cyclic quadrilateral formed by the altitudes.)

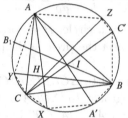

Because A_1 is the midpoint of arc BC, area $\Delta BA_1C \geq$ area ΔBXC.

So area hexagon $AC_1BA_1CB_1 \geq$ area hexagon $AZBXCY$

$$= 2 \times \text{area } (\Delta BHC + \Delta CHA + \Delta AHB)$$
$$= 2 \times \text{area } \Delta ABC.$$

67. Let $AB = c, AC = b, BC = a$ and $AD = h$, the circle inscribed in ΔABD by C_1 and that inscribed in ΔADC by C_2. Let O_1, O_2 be the centres of C_1 and C_2, respectively, E and F the points of contact of C_1 with AB and AD, respectively, and G and M the points of contact of C_2 with AD and AC, respectively. Let O_1N be perpendicular to AC and O_2P perpendicular to NO_1.

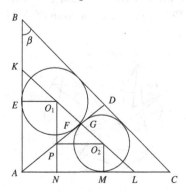

Then

$$O_1N = EA = AF = h - r,$$
$$O_1P = O_1N - PN = O_1N - O_2M = h - r - R,$$
$$O_2P = MN = AM - AN = AG - r = h - R - r.$$

Hence $PO_1 = PO_2$ and thus $\angle O_1 O_2 P = 45°$. Therefore $\angle O_2 LM = 45°$ and thus $ML = O_2M = R$. Consequently $AL = AM + ML = AG + R = h - R + R = h$. Similarly $AK = h$. Hence

$$\frac{S}{T} = \frac{ah}{h^2} = \frac{a}{h} = \frac{a^2}{ah} = \frac{b^2 + c^2}{bc} \geq 2.$$

68. Let $\triangle ABC$ be inscribed in a circle. H its orthocentre, I its incentre. Extend each altitude to meet the circle in the points A'', B'', C'', in the natural way.

Then $\triangle CA''B$ is congruent to $\triangle CHB(\angle A''CB = \angle BCH = 90° - B)$, similarly $\triangle CB''A$ is congruent to $\triangle CHA$ and $\triangle BC''A$ is congruent to $\triangle BHA$. So the hexagon $AC''BA''CB''$ has area double the area of $\triangle ABC$. (*)

$$\angle B'XA = \angle B'C'A + \angle XAC'$$
$$= \angle B'C'A + \angle XAB + \angle C'AB$$
$$= \frac{1}{2}B + \frac{1}{2}A + \frac{1}{2}C = 90°$$

so $A'A$ is an altitude to $\triangle A'B'C'$. Then by statement (*) above, the area of the hexagon $B'CA'BC'A$ is twice the area of $\triangle B'C'A'$.

But the area $\triangle CA''B \leq$ area $\triangle CA'B$, as A' is the midpoint of the arc CB, and similarly area $\triangle CB''A \leq$ area $\triangle CB'A$ and area $\triangle BC''A \leq$ area $\triangle BC'A$, so the area of the hexagon $AC''BA''CB'' \leq$ the area of the hexagon $B'CA'BC'A$, or twice the area of $\triangle ABC \leq$ twice the area of $\triangle A'B'C'$, as required.

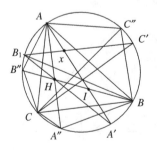

69. As usual we let $A, B,$ and C denote the measure of the corresponding angles, and let a, b and c be the lengths of opposite sides. Referring to the description of L_A we have

$$\angle AMN = \angle AHN \text{(angles subtended by the same arc are equal)}$$
$$= 90° - \angle HAC \text{(a diameter subtends a right angle)} = C.$$

Similarly $\angle ANM = B$. The triangle AMN is therefore similar to the triangle ABC. In particular it is acute angled. It follows that the line L_A meets MN between M and N, and so it meets BC at a point P between B and C. Hence

$$\angle BAP = 90° - \angle AMN = 90° - C \quad \text{and}$$
$$\angle CAP = 90° - \angle ANM = 90° - B.$$

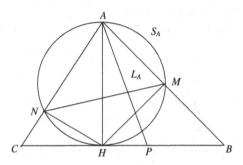

Let O be the circumcentre of triangle ABC. Then $\angle BAO = \angle ABO$, $\angle CAO = \angle ACO$ and $\angle CBO = \angle BCO$. By adding together all the angles we deduce that $\angle BAO + C = 90°$. Hence $\angle BAO = 90° - C$. But then AO must coincide with L_A. Similarly O must lie on L_B and L_C.

70. The inradius of a triangle is the ratio of its area to its semiperimeter, hence we obtain

$$\frac{\frac{1}{2}xa}{\frac{1}{2}(a + x + \sqrt{x^2 + a^2})} = \frac{\frac{1}{2}(b - x)a}{\frac{1}{2}(c + b - x + \sqrt{x^2 + a^2})},$$

$$x(c + b - x) + x\sqrt{x^2 + a^2} = (b - x)(a + x) + (b - x)\sqrt{x^2 + a^2},$$

$$xc + xa - ab = (b - 2x)\sqrt{x^2 + a^2},$$

$$0 = (b - 2x)^2(x^2 + a^2) - (xc + xa - ab)^2$$

$$= 4x^4 - 4bx^3 + (3a^2 + b^2 - c^2 - 2ac)x^2$$

$$+ (2abc - 2a^2b)x$$

$$= 4x^4 - 4bx^3 + (2a^2 - 2ac)x^2$$
$$+ (2ac - 2a^2)bx$$
$$= 2x(x - b)[2x^2 + (a^2 - ac)],$$
$$\therefore \quad x = 0, \ b \quad \text{or} \quad \pm\sqrt{(ac - a^2)}/2.$$

Of these, only $x = \sqrt{(ac - a^2)}/2$ provides a valid geometric solution, and we can be sure that $0 < x < b$ (as required) because
$$ac - a^2 = a(c - a) > 0 \quad \text{and}$$
$$ac - a^2 = b^2 - c(c - a) < b^2.$$

71. Let CC' be an altitude. Since $AH = R$, the circumradius, we have
$$AC' = R\sin B \text{ and hence } CC' = R\sin B \tan A. \tag{1}$$
   From the fact that $\sin A = BC/2R$ and that $CC' = BC\sin B$ we have $CC' = 2R\sin A\sin B$. $\hspace{3cm}$ (2)
   Equations (1) and (2) imply that $2\sin A = \tan A$, so A must be $60°$.

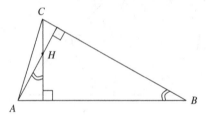

3. (revisited) In the diagram below, $\triangle CA''B''$, the circumcentre O lies **inside** the triangle. Since $\angle A''OB'' < 180°$, $\angle C''A''B'' < 90°$. A similar result follows for all other angles of $\triangle AB''C''$. Hence this triangle is acute.

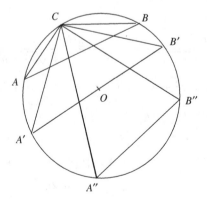

A similar argument shows that $\angle CAB > 90°$ and so $\triangle ABC$ is obtuse.

We can now easily prove that an acute angle triangle contains its circumcircle. Suppose the triangle is acute but the circumcircle lies **outside** the triangle. Then by the argument of the last paragraph, the triangle is obtuse. This contradicts the original assumption about the triangle.

So we have a triangle is acute if and only if its circumcentre lies inside the triangle.

In a similar way we can prove that a necessary and sufficient condition for a triangle to be obtuse is that its circumcircle is outside the triangle. What are necessary and sufficient conditions for a triangle to be right angled?

6. (revisited) Let CD be a diameter of the circumcircle of the equilateral triangle ABC.

Since $ADBC'$ is a cyclic quadrilateral, $\angle DAC' + \angle DBC' = 180°$. But $\angle DAC' = \angle DAC + \alpha = 90° + \alpha$. Similarly $\angle DBC' = 90° - \beta$. Hence $90° + \alpha + 90° - \beta = 180°$. So $\alpha = \beta$.

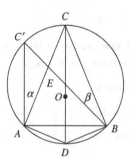

In $\triangle$'s $AC'E$, BCE we have $\alpha = \beta$ and $\angle AEC' = \angle BEC$ (vertically opposite angles). Hence $\angle AC'B = \angle ACB\,(= 60°)$.

(Does this proof generalise to triangles other than equilateral triangles?)

18. (revisited) In the diagram of the solution of Exercise 18, draw CO extended to C' on the circumcircle. The proof that $\frac{b}{\sin B} = 2R$ follows as in Exercise 15 using $\triangle CC'A$. For $\frac{a}{\sin A} = 2R$, use $\triangle CC'B$.

35. (b) (revisited) Clearly there are problems with a $45°, 45°, 90°$ triangle. But this is covered by part (a).

36. If and only if the line XY is perpendicular to the other two lines. (Suppose XY is drawn as in Figure 2.7. Then the circle in region II touches XY **above** the midpoint of XY while the circle in region VI touches XY **below** the midpoint.)

2.9. Glossary

The numbers in brackets refer to the pages where the term is defined.

altitude (48): of a triangle, the line from a vertex to the opposite side, perpendicular to the opposite side.

centroid (48): of a triangle, the point of intersection of the three medians of the triangle.

circumcentre (42): centre of the circumcircle.

circumcircle (42): circle through the vertices of a triangle actually the term is also used for the circle through the vertices of any figure.

circumradius (42): radius of a circumcircle.

cyclic quadrilateral (44): four-sided figure whose vertices lie on a common circle.

Euler line (57): line joining the orthocentre, the centroid and the circumcentre of a circle.

excentre (50): centre of an excircle.

excircle (49): circle such that the sides of a triangle produced, are tangents to the circle with the circle outside the triangle.

exradius (50): radius of an excircle.

incentre (46): centre of an incircle.

incircle (46): the largest circle that will fit inside a triangle with all three sides as tangents.

inradius (46): centre of an incircle.

medial triangle (56): triangle whose vertices are the midpoints of the sides of another triangle.

median (47): line from a vertex to the midpoint of the opposite side.

nine point circle (56): for a given triangle, the circle through the midpoints of the sides, the feet of the altitudes, and the three midpoints of lines from the orthocentre to the vertices of a triangle.

orthic triangle (48): the triangle whose vertices are the feet of the three altitudes of a triangle.

orthocentre (48): the point of intersection of the three altitudes of a triangle.

Chapter 3

Solving Problems

3.1. Introduction

In this chapter I want to first say what I think mathematics is, that is, I want to explain the steps that some mathematicians go through when they are doing research. From here I want to show the connection between these steps and what anyone does when they are attempting to solve a problem. Along the way, in fact right from the start, I want to "solve a problem" so that I can illustrate the things I'm trying to convey.

But the point of this chapter is to give you some ideas that you can use when you are stuck on a problem — when you have no idea what to do next.

Hopefully over time, this will be the most useful chapter both of this book and of *First Step*. Hopefully you will come back again and again to look at what is being suggested here.

Having said that, I think that you should probably read through this chapter fairly quickly, taking note of the main ideas, and remembering that there is some useful material here that you can use later.

3.2. A Problem to Solve

Look at Figure 3.1. Is it possible to put the numbers 1, 2, 3, 4, 5, 6, one per circle, so that the three numbers on each side of the equilateral triangle have the same sum? If it can be done, in how many ways can it be done? If it can't be done, why can't it be done?

I'm going to assume that you can't do this problem. In fact, even harder, I'm going to assume that I can't do this problem. I'm going to assume that I haven't seen this problem before. So what would I do first? I think that I would play around a bit; put the numbers in the circles and hope that I got

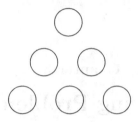

Figure 3.1.   An arrangement of six circles.

some inspiration. So why don't you do that and see what you come up with? Oh and it's going to be useful to have a word for one of the arrangements that would satisfy the problem, if such an arrangement existed. So I'll call such an arrangement a **snark**, after the mystical creature in Lewis Carroll's famous poem "The Hunting of the Snark".

Exercises

1. Put the numbers in the circles, anywhere so long as it's one number per circle. Can you find a way of making the three numbers on the left side of the triangle add to the same thing as the three numbers on the other two sides?
2. If you do get success in the last exercise show that you have all possible snarks. If you don't get success, show that there are no snarks.
3. What ideas did you come up with in this search for snarks?

    Now in the solution to Exercise 2 above, I have listed six possible snarks. But you should be able to see that two of these are just the same as two of the others, so in fact there are only four *different* answers.

    Unfortunately I haven't solved the six circles problem completely yet though. I still have to show that these are the only answers possible.

Exercises

4. Are there only four snarks or can you find more?
   However many answers you find, can you show conclusively that there are no more? How could you do this?
5. What properties do the answers have? How do the different answers differ? How can we use these differences?

    Are the sums of the numbers on each side restricted in some way? Can we have arbitrarily large side sums? Surely there has to be some restriction?

After all, we shouldn't be able to get a sum bigger than $4 + 5 + 6 = 15$ or smaller than $1 + 2 + 3$?

Now 1 has to be somewhere in the answer. What is the largest sum that I can get on the side that contains 1? Isn't that $1 + 5 + 6 = 12$? So I can't get a side sum greater than 12.

Exercises

6. Show the upper limit of the value of the side sum is 12.
7. Show that there is only one snark with a side sum of 9.
   Is the same true for each of 10, 11 and 12?

I have deliberately led you through a specific solution to the six circles problem. There is at least one quite different approach to this but I won't give it to you right now. Instead, now that I have solved this specific problem I want to look at how some mathematicians attack research problems.

3.3. Mathematics: What is it?

One way to answer this question is to reach for the nearest dictionary and find a definition of "mathematics". But I'm too lazy to even look up the word on the web. Besides it wouldn't help. The result would be, at best, a one-dimensional view of something that is infinitely-dimensional.

Let me start off slowly to try to explain what I think mathematics is. I guess that mathematics is a way of looking at the world and trying to sort out some of its problems. Mathematics seems to have presently accumulated an enormous number of facts, ideas and theorems. But, despite the way that maths is still taught in many classrooms in the world, it is more than a collection of algorithms that, for some unknown reason appear to *have* to be known. There is also a creative side to the subject — a place where new maths suddenly appears, sometimes even miraculously. And the thing that mathematicians have tried to suppress from the general public: that it is created by human intervention.

In my own research career it seemed to me that there was a rough structure around the creation of new ideas and results. Writing this down in some ways doesn't help because any structure that can be put on a page will be inadequate for the grand task it is written down for. Nevertheless, it is the only way I can think of to communicate it so here is my first attempt at a structure for the creative side of mathematics.

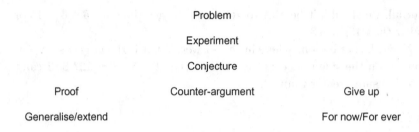

Figure 3.2.   An attempt to describe the structure of the creative process in mathematics.

What you see in Figure 3.2 are the key points of a process. I'll try to weave these points together and show how they interact. But right at the start it is necessary to say that their interactions are in no way linear and often appear to be quite random.

To me there is no mathematics without a ***problem***. Mathematics exists to solve problems. I don't see how you can do maths unless you are working on a problem. I also want to say that by problem, here, I mean something that it is not immediately obvious how to solve. OK I may have some thoughts on the matter, some area of maths to look at for a solution, even some theorems that I might think that I might employ in the solution, but I can't just sit down and write out a solution. These are not exercises that I am dealing with where the method is known; these are genuine problems. Clearly if they are research problems, then nobody yet knows how to solve them.

A common combinatorial approach to solving problems is to ***experiment*** with the concept in hand. This may involve using a computer to generate examples that can be counted or examined in some way. It may involve generating examples by hand. But the aim of the exercise is to get to know the problem I am dealing with in more detail so that I can understand the problem better and form some idea as to how a whole class of objects might behave. And when I have that idea I have a ***conjecture*** as to what might be going on. A conjecture is just a guess as to what might happen, but a guess grounded by the examples I have just produced and by my overall mathematical experience and intuition.

If I am lucky the conjecture will be true and I can prove that it is true. But I won't know this beforehand, so it may well be false. In this indecisive position I'll try to prove it but a proof may not come either because there isn't one or I don't have the mathematical machinery to prove it. So I'll be caught in a wasteland where I alternate between looking for a proof and looking for a counter-argument or a counter-example. By a **counter-example** here, I mean an example that shows that the conjecture is wrong. While I think about it, in my experience, most of the conjectures that I have had have been false! Only perhaps 1% of conjectures have proved to be true eventually. But posing these conjectures is a good way to make progress. They focus the mind and suggest ways to make progress.

Hopefully, any difficulty with a proof will lead to an idea for a counter-example and any difficulty in constructing a counter-example will give more fuel for the construction of a proof.

But what about the **give up** part of Figure 3.2? All dead mathematicians clearly have given up on everything that they were working on. Many live mathematicians have also given up, but on only some of the problems that they would like to solve. They can't see how to marshall their mathematical armoury in such a way as to crack a problem. They might hope that they will get back to the problem at some point but there is no guarantee that they will; so sometimes problems are given up for ever. That's just the way it is.

But giving up **for now**, is another kettle of fish. This is actually a very good tool for solving problems. Many times, after I, or the group of people I was working with at the time, had been working hard on a problem, exhaustion, or lack of ideas made me (us) give up. But just because I had given up it didn't mean that my brain had. Somewhere deep in the bowels of my brain there is someone who is always at work. When I least expect it, that someone sends a virtual text message to the upper levels of my brain. This is an idea that gets over whatever temporary impasse that had caused me to give up and suggests a new way forward.

These, sometimes brilliant, ideas are a matter of public record. Hamilton, a famous Irish mathematician, had worked on quaternions for a long while. Suddenly, when he was walking along the banks of a canal with his wife, the solution suddenly came to him. He was so excited that he carved the basic idea on the first bridge that he came too. This carving may have been a way to remember what he had just discovered in case he had forgotten it by the time he got home. But Poincaré needed no such *aide de memoire* when he solved his problem. Again, he had worked on it for a while and it was only as he got on a bus to go home that the brilliant thought popped into his head.

So sometimes when you are looking for ideas it's a good idea to stop looking for ideas. Just give up. If you have been working hard enough to solve a problem, something in your subconscious will continue to work and, when you least expect it, may well text you a solution.

The last part of Figure 3.2 talks about generalise and extend. To *generalise* a problem is to find another problem that includes the one you have been looking at. So if I can solve this generalisation I will get a much more powerful result than I would have got by solving the first problem. An *extension* on the other hand, is a similar problem to the one that I have just solved. By tweaking the original problem, by changing it in some way, I get another problem that may well be worth solving.

Mathematics is continually being extended, not just by solving problems, but also by solving generalisations and extensions of problems.

3.4. Back to Six Circles

As far as it went, there is virtually no difference between solving the six circles problem and doing mathematical research as it was described in the last section. In both cases we started off with a **problem**. Then we **experimented** to see what we could find out about the problem (and also to understand the problem better). Next we had a **conjecture** that there were only four answers. With a bit of work, we eventually found a **proof** of the conjecture.

The structure of the two processes, research and solving problems, is much the same, as is the thinking along the way.

Now I haven't talked about generalisations or extensions in the context of the six circles problem, but I can, and will. If I might want a *generalisation*, I might ask for all sets of 6 different numbers that can be put into the circles so that the side sums are the same. If I can find this most general set of 6 numbers, then I only have to check to see if $\{1, 2, 3, 4, 5, 6\}$ satisfies these general conditions. If it does, then I get the solution I wanted to the original problem. Clearly, the general situation is more powerful than the particular one that I started with.

As for *extensions* I need to change the problem slightly. Suppose that I put three circles on the sides of a square. This will give the eight circle problem and I can see if it is possible to put the numbers from 1 to 8 in these squares so that the three numbers on each side of the square add up to the same value.

Note that if I can do this, I don't necessarily get a solution of the six circles problem, but it is still an interesting problem and may lead to an interesting result.

Exercises

8. Invent three more generalisations of the six circles problem.
9. Invent three more extensions of the six circles problem.
10. Let a set of six numbers be *nice* if the numbers can be put into the six circles so that the sums of the three numbers on any of the three sides of the equilateral triangle are the same. Describe nice sets. Prove that what you have conjectured is true.
11. In what other problems in this book did you need to (i) experiment; or (ii) conjecture?
    Did you ever find a counter-example?
    Did you ever give up?
    Now that you know about generalisations and extensions, are there any problems that you have solved and you can now generalise or extend?

3.5. More on Research Methods

Perhaps not surprisingly there is more to mathematical research than the skeleton of Figure 3.2. In fact the same is true for solving problems. In this section I want to say something about heuristics. Now *heuristics* are ideas that are worth trying. They won't always get you to a solution of your problem but they may well help. They are worth a try anyway. So let me look at a few of the more commonly used ones. Here is a list *of heuristics*, in no particular order, to start off with.

   (i) think of a similar problem
  (ii) guess and check
 (iii) draw a diagram
 (iv) try some algebra
  (v) act it out
 (vi) make a systematic list
 (vii) look for patterns
(viii) consider a simpler problem
 (ix) consider an extreme case.

   (i) *Think of a similar problem*: This, of course, is the first thing that you think about when you are doing an exam. "What sort of question is this?" "Is it something like the one I was revising?" And in trying to solve a problem, you should always start by asking "Have I seen anything like this before?" It is just possible that some problem that you have done recently is similar in some way to the one you are trying to solve. There may be something in the solution of the earlier problem that will help you in the current one.

An example of this is Exercise 46 of Chapter 2, which depends to some extent on Exercise 27. It is very helpful in trying to find the centre of the circle through the midpoints of the sides of a right angled triangle, if you know that pairs of these midpoints form lines that are parallel to the sides of the original triangle.

The same thing is true for Exercise 68 of Chapter 2. What needs to be done here is very similar to what has been done in Exercise 66. Indeed many of the exercises of this book rely on proof methods or ideas established in earlier exercises. So you should always try to see if there is something helpful in some preceding exercise.

(ii) *Guess and check*: Isn't this what you did when you experimented with the six circles problems to get an idea of whether or not there were any snarks? I suspect that just plugging in numbers (guessing), followed by checking that the side sums were or were not equal, enabled you to make progress there. Maybe the same thing could be said of many other problems that you have met recently.

In fact, when you think about the research process I set down above, pretty much everything in mathematics is guess and check. Much of the early work on a problem is guessing a good conjecture. Then you check it by finding a proof or a counter-example.

That makes guess and check sound very sophisticated, but in many ways it is the most basic of all heuristics. All you do is to plug in a few numbers, draw a few diagrams or whatever and check that something you have done makes sense. When all else is failing, just guess and then check your guess. Hopefully by eliminating useless guesses, you'll come up with something worth pursuing. So don't forget it as a strategy when you can't see any other way to go.

(iii) *Draw a diagram*: There is very little geometry that can be done without using a diagram. It helps you to understand what the problem is asking and it helps you to guess how things might work. A big diagram, poky little ones often lead to bad intuition, is a great aid in getting started in most geometrical problems.

But diagrams help in many other problems. I still remember when I went to hear my first lecture on finite simple groups, an aspect of a thing called abstract algebra, being surprised by the fact that in an algebraic topic, the experts were using **diagrams**! These diagrams clearly helped them to see what was going on, to understand quite abstract ideas, and to communicate the work to each other and to lesser mortals like me.

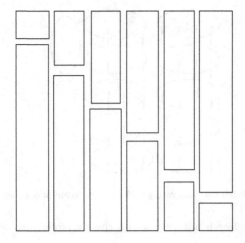

Figure 3.3.   The sum of the first six numbers is a half of 6 × 7.

One of the easier ways to sum the first six numbers, for example, is to put the numbers into rectangle form, see Figure 3.3. There the rectangles shown have dimensions $1 \times 1, 1 \times 2, 1 \times 3, 1 \times 4, 1 \times 5$ and 1×6. Two lots of these rectangles cover another rectangle whose dimensions are 6 × 7. So twice the sum of the numbers 1 to 6 "=" the area of the rectangles = 6 × 7. This tells us that the sum of the numbers is (6 × 7)/2.

In a similar way we can add the numbers from 1 to 100 and, in fact, we can add any numbers in arithmetic progression this way.

Some problems that you might think are only open to an attack by algebra, are open to solution by diagram, and so can be solved by quite young students. One such problem is the Farmyard Problem. There are 15 animals, turkeys and pigs, in the farmyard. Altogether there are 50 legs. How many turkeys are there? You can solve this using a diagram and guess and check.

I've put rings around groups of four legs to represent the pigs. In Figure 3.4, there are 12 pigs and two legs left over. One turkey equals two legs and so that only gives us 13 animals. By gradually reducing the number of pigs (and consequently increasing the number of turkeys), I'll eventually get the complete answer.

(iv) *Use algebra:* Now it's not clear to me that this is in the same league as the other heuristics but it is certainly something that is well worth trying. One of the advantages of algebra is that once you have set up an equation, say, then you get a long way without having to think too much about the problem. The algebra itself takes over and all you have to do is to follow it.

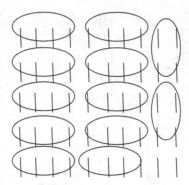

Figure 3.4.   Solving the farmyard problem by diagram and guess and check.

Maybe that's a bit harsh but let's set up the six circles problem by algebra
to see what we get.

Suppose that the letters p, q, r, s, t, u are placed consecutively in the
circles and that p, r and t are in the corner circles of a snark. Further let s
be the side sum. Then

$$3s = (p + q + r) + (r + s + t) + (t + u + p).$$

So

$$3s = p + q + r + s + t + u + (p + r + t) = 21 + (p + r + t).$$

This is because the numbers from 1 to 6, in any order, add up to 21. But
what is the biggest and smallest that $p + r + t$ can be. Surely $1 + 2 + 3 \leq$
$p + r + t \leq 4 + 5 + 6$. And this gives us

$$27 \leq 3s \leq 36$$

Straight away we see that the side sum must be anything between 9 and 12.
From here it's not hard to find all four snarks.

Maybe I did have to do some thinking there, but it is certainly true that
if you can set up some sensible algebra it often makes the problem easier to
solve.

(v) *Act it out*: One technique that is quite useful in a number of
problems is to actually put yourself in the problem. For instance, look at
the frogs problem (see http://www2.nzmaths.co.nz/frames/brightsparks/
frogs.asp?applet). Here we have three brown frogs on three different lily
pads and three green frogs on three other lily pads. All the lily pads are
in a line and there is one empty lily pad between the two groups of frogs.

Now the green frogs want to get to where the brown frogs are and vice versa. Frogs can move by going to an adjacent lily pad that is empty or by jumping over a frog onto an empty lily pad on the other side of the jumped frog. One final rule, frogs can't go back. If they start moving to the left, they have to keep moving in that direction. And the problem is, can the green and brown frogs swap places?

A very good strategy here is to put seven chairs in a row to represent the lily pads, and get two groups of three students to sit on the chairs and **act the problem out**. What are the advantages of doing this?

First of all it has all the advantages of a (three-dimensional) diagram. It helps you see what is going on and it helps you remember what you have done. It's also easy to see when you have made a wrong move and to diagnose it was wrong. Second, it helps you to see ideas. Third, it helps you to see patterns and enables you to make obvious generalisations and extensions. What happens if we have four frogs on each side or five or n? What happens if we have three frogs on one side and five on the other? How about m and n frogs on different sides? And fourth, it seems to be more fun this way. And don't underestimate the importance of enjoying what you are doing in maths.

There are several "crossing the river" problems that can be solved in a similar way (if you don't know about them you can look them up on the web by looking for "crossing the river problems"). Take the man who wants to get his fox, goose and some grain across a river but his boat is too small to take more than one at a time. But he can't leave the fox and the goose because the fox will eat the goose and he can't leave the goose and the hay because the goose will eat the hay. How can he get all of his belongings across the river?

But "act it out" is also a good strategy to understand game problems. In games with two players it's often better to have another person there "competing against" you. They will often see things from a completely different perspective and bring ideas to the game that you may not have thought of.

For instance, think about the Nim-like "21 Game". Here there are 21 pieces on the table and each player takes a turn to remove some. Both players have to take 1, or 2 pieces from the table. The winner is the person who takes away the last piece. Does the first or second player win? I'm assuming that both players play optimally, that is, as well as is possible.

Even if you don't have another person on hand to tackle such a game, it is probably better to play this game against yourself with actual pieces.

Usually this is much better than just thinking about it. Somehow using pieces helps you see obvious errors that you might otherwise make.

One final note here that goes beyond "acting it out", is the value of drawing other people into conversations about a problem. If you are lucky they will certainly look at things from a different angle to you and that will help you to think in new directions. But even if they know nothing about the problem, or even about mathematics, sometimes just talking out loud can help you move forward with some new ideas. So don't feel silly if you talk about a problem to your favourite pet. Surprising as it may seem, this is likely to help you make progress.

(vi) *Make a systematic list*: In answering Exercise 7 on the six circles problem, you could use a systematic list. I used it to show that 9 can only be produced in three ways using three of the numbers 1 to 6 and using no number more than once. To do this I first said let me use the number 6. Now I can't use 5, 4, or 3 now because they will give me sums bigger than 9. But I can use a 2. Then I have to use 1 as my final number. So the only sum of 9 I can get using a 6 is $6 + 2 + 1$.

Now I won't use 6 any more. Let 5 be the biggest number I'll use. At this point I can't use a 4 as that gives me 9 and I still have a number to go. So what if I chose 3? Then I could finish off with 1. But what if I didn't choose 3? I could use 2, but then I would need a 2 to complete the 9 and I know that I can't use the same number twice. So the only way to make 9 with 5 as the highest number is $5 + 3 + 1$.

Now let the biggest number be 4. Suppose that I take 3 next. Then I will have to finish with 2. But what if I didn't take 3 but took 2. Then I would need 3 to finish. But I can't do that as I have considered the use of 3. So what if I took 1. Then I'd need a 4 to make up the 9. That's no good as I've already used 4. So the only way to get 9 if 4 is the biggest number is $4 + 3 + 2$.

Then suppose that 3 is the biggest number I can use. But that's a problem. I can't use 3 any more so let me try 2. Now I can't get 9. And the same thing happens if I take 1 after the 3. So there are no ways to get 9 using 3 as the biggest number.

Systematically then, I've shown that there are only three ways to get 9 given the conditions imposed by the problem. Now you can see how I got the three sums in Exercise 7.

There are many problems where a systematic list is a useful tool in getting a solution. Commonly these are in problems in Number Theory or Combinatorics.

Another type of systematic list is used to cover all possible cases in a problem. Sometimes a problem naturally falls into a few cases that you can solve or tackle separately. This was the approach we took in Section 1.7 when we looked at derangements. First we looked at the case where two letters were "swapped", that is, where letter A went into box B and where letter B went into box A. Then we looked at the case where this didn't happen.

If you come on a problem that separates into cases, try to do the easiest one first. This will often give you ideas on how to tackle some of the other cases. But be sure that the cases are disjoint and do cover the whole problem.

(vii) *Look for patterns*: In some sense, all of mathematics is about looking for, and finding, patterns. But there are some problems that lend themselves more to this than others. A number of these problems appeared in *First Step*. For example, the jug problem and the postage stamp problem. The same thing occurs in the 21 Game I mentioned in "act it out" above.

(viii) *Consider a simpler problem*: This is often a way to look for patterns. If the 21 Game, for example, is too hard to analyse, then you might try the "3 Game". If there are only three pieces on the table, it ought to be easier to see what is happening and how one or the other player might play to win. From there you might work up to the "4 Game" and so on until you have enough evidence to see a pattern and then decide what is the best strategy for the 21 Game (and maybe even for the "n Game").

Similarly if you want to find a way of swapping the two sets of three frogs, you might not do better than to see what happens first for $n = 1$ and then 2.

You may remember that I prepared for the nine point circle by looking at the simpler cases of the equilateral triangle and the right angled triangle.

Exploring and investigating simpler cases is a very useful heuristic to have up your sleeve.

(ix) *Consider an extreme case*: This is a variation on "consider a simpler problem", but here we are looking to test the limits of what we are doing. For instance, suppose that I want to find the range of values that the side sums can have in the six circles problem. Then I might ask "can I find a snark that has a side sum of 100?" This is clearly extreme and so extreme in this case as to be useless. But follow that by replacing 100 by 99 and so on until the potential side sum is no longer extreme. In the process I might see some way to limit the values of the side sum.

"Considering an extreme case" is a way of testing the limits of the problems to see what can happen and what can't happen. It's aim, and the aim of all heuristics, is not to solve the problem but to tell you more about it so that, by narrowing down the options you can eventually solve the problem.

Exercises

12. Show that a set of nice numbers has four different snarks if and only if the numbers are in arithmetic progression.
13. Go through the heuristics of this chapter and find as many problems or exercises as you can where the heuristics might be applied.
14. By how many different methods can you solve the Farmyard Problem?
15. Show how to solve the frogs problem for n frogs on each side.
    How many moves does it take?
    Which of the things in the last section were helpful to you here? Why? And in what part of the solution?
16. Solve the 21 Game. In other words find out what are the best strategies to use and who will always win — the first or second player.
    Solve the n Game.
    Which of the things in the last section were helpful to you here? Why? And in what part of the solution?

3.6. Georg Pólya

There are many very useful heuristics that I have omitted from the list in the last section, but you can find a lot more in the book by Georg Pólya called "*How To Solve It*". This is a valuable book that can be found in most libraries and is easily bought via the web. Apart from providing heuristics it has a lot of interesting problems and sets out the first real attempt to think about how to approach the solving of problems.

Pólya's problem solving model has just four steps. These are:
  (i) Understand the problem;
 (ii) Devise a plan;
(iii) Carry out the plan;
(iv) Look back.

The steps in the process of mathematical discovery are very much what Pólya had in mind when he listed the four steps above. Pólya was an important mathematician who was aiming to list the problem solving steps for high school students and the order in which they might be carried out.

In "Understanding the Problem", he emphasised the fundamental need to know what the problem was asking before trying to make any further progress. Of course it is necessary to solve the problem in hand rather than one that you might have thought up that appears to be the one in hand. This is clearly the case in examinations or tests. So before you get too far down the problem solving track you need to satisfy yourself that you have understood the problem and know what it is asking.

This first step involves reading the problem several times, not just at the start but throughout the process so as to not stray from the track. It also involves the experimentation that I suggested in Section 3.1. It might go as far as getting a first conjecture.

Then it is important to have a plan of where you hope to go. You might first plan to get to understand the problem by doing a few simple examples. That might be followed by finding a conjecture. But then, how are you going to prove the conjecture is true? What mathematical ideas might help you? Algebra? Can I set up a Proof by Contradiction or even by Mathematical Induction[a]? When have these types of proof worked before? Will they work here? So how do they fit into my plan?

At first it might be useful to actually say what that plan is but as you get more experienced you will have a better idea of what you are trying to do and probably won't need to write it down.

After having given some thought as to what your plan might look like, then you are set to carry the plan out. This won't necessarily go smoothly. Maybe, say, you had thought to use a Proof by Contradiction, and you can't set it up as you thought you might. Then you just have to go back and plan again. It may be too that you don't know some piece of mathematics that is fundamental to a solution. That is a good time to look up some books or the internet or ask someone who might know.

But if your plan works and you solve your problem, then it is time to look back. First it is worth checking that the problem you have just solved is the problem you were actually asked to solve. If not, it's back to the drawing board. If it is, then you might check that you have solved it correctly. Did you make any errors? If not, perhaps there is a better way to solve it. (It's amazing the number of times that a mathematician has produced a theorem and when writing it up to publish it in a journal, finds a better proof.) After that, you can think about where this problem that you have solved might lead. Are there any other interesting things that might come out of this?

---

[a]See Chapter 6 of *First Step*.

Can you generalise or extend the problem in any way? Follow up any leads to see where they take you.

And after that you've pretty much done as much as you can and it's on to the next problem.

3.7. Asking Questions

So far, I think that from the above discussion, I have shown that mathematicians have three strings to their bow when they are trying to do research. First of all they know as much maths as they can. Indeed they are always learning more so that they can tackle more. That is one of the reasons that they go to conferences, talk to colleagues and read journals.

Second, they have a structure that they work through; something like Figure 3.1 or the Pólya four-step approach. The more research they do the more implicit this structure is and they generally don't think about it.

Then they have strategies that they can call on — heuristics as we have called them above. These are not things like formulae or equations but ideas of approaches that might work. Heuristics may put mathematicians onto the right track but they may also be useless. This is the difference between problems where there is no obvious way forward and exercises where there is. Often mathematicians may try all the tricks at their disposal and still not solve what they are working on. Some problems have been around for hundreds of years and are still not solved (look up Goldbach's Conjecture, for example).

At this "stuck" point, it is time to ask questions. Indeed it is almost always time to ask questions. If anyone else is around you can ask them. If not, ask yourself. In fact, when I am stuck, really stuck, I often wish that I had a teacher person at hand who knew the answer and could tell me what I need to do next. If they wouldn't actually tell me the answer it would be good if they could at least tell me if I am on track. Should I keep going in this direction? Am I close to the solution? Or should I try something different? Suppose this teacher person would only give me a hint. What might that hint be? How might they phrase that hint in terms of a question?

But what questions are useful? This is actually a very difficult question and the answer depends very much on where you are in the problem solving process and what the problem is that you are working on. In the early stages of solving a problem you might want to use a heuristic and ask if you have seen a problem like this before. You might also want to ask questions that ensure that you have understood the problem. In the six circles problem you might want to just put in some values and then ask if the arrangement you have created satisfies what the problem asks. Have I used all of the numbers? Do all of the sides add up to the same number?

If you are satisfied at that point, then you might want to work forward towards a plan. If I knew "this" and "this" would that get me through to the end? What are "this" and "this"? For example, in the six circles problem, I might see that if I knew the values that the side sums had, and if I had a way to show how many of these produced the arrangements I wanted, then I might be finished. But that leads to how can I restrict the side sums? And that might lead to looking at extreme cases.

So I have a plan and I'm carrying it out, but somehow I'm stuck again. Note that if I'm not stuck then I don't have to worry about asking useful questions. So I'm stuck. Why could I be stuck? Is it because I don't know some maths that would be useful? Or is it I'm missing the right strategy? Or have I just got a bad plan? At this point I might want to ask all of these questions and even consider "have I been going down this path for too long?" Maybe it's time for a new plan or a new idea.

On the other hand, I may have got through my plan and have found the answer or answers to the problem. At this point I certainly want to check that I have solved the problem that was actually asked. And also I want to be sure that I have indeed solved it — that all of the steps are correct. Have I made any mistakes? A positive answer to this would probably put me back to the start again. A negative answer will suggest that I should try to extend or generalise. How could I do either of these things?

It's not really possible to give a complete flavour of what questions to ask and when but it is generally believed that it is more important to find the right questions than almost anything else. If you can use questions to shave away the layers of the problem you will lead yourself to the answer. So it is important to learn to ask questions because then the right answers will follow.

Exercises

17. Solve this problem.

    Three aliens and three humans want to get from the moon they are on to another moon close at hand. They must cross between the moons in a shuttle that will only hold two creatures at a time. If the aliens ever outnumber the humans, they will do unspeakable things to them. How can all three aliens and all three humans cross to the new moon safely?

  (i) What of the steps of Figure 3.2 did you use?

 (ii) What steps of Polya's four step process did you use?

(iii) What heuristics did you use?

 (v) What mathematics did you use?

(vi) Write down the questions that you asked yourself.

18. Repeat the last exercise with the "400 Problem". Essentially, you have these two subtractions:

$$\begin{array}{ccc} 4 & 0 & 0 \\ - \ a & b & 4 \\ \hline . & . & . \end{array} \qquad\qquad \begin{array}{ccc} 4 & a & b \\ - \ 4 & 0 & 0 \\ \hline . & . & . \end{array}$$

If you know the answers are the same for both subtractions, what are the values of a and b?

3.8. Solutions

1. I can't answer this question for you, but here are a few random guesses.

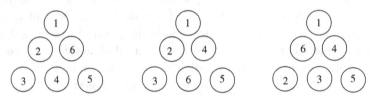

2. I managed to find some arrangements that did what the problem asked. These are shown below.

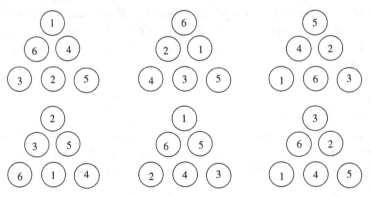

3. I don't know what you did but you might have tried systematically to put the numbers in all possible ways. You probably used lots of diagrams with six circles in them. Was there any way that you found to be systematic? What about using algebra in some way? Did you have any other ideas on how to tackle the problem?

4. Yes there are only four snarks, but I'll hold back on showing you why for the moment.

5. Since the problem places importance on the sums of the numbers on each side, this property might be worth considering. Why are the only side sums 9, 10, 11 and 12?

6. What is the smallest side sum for a side containing 6?

7. Nine can only (see "systematic list" in Section 3.4) be written as the sum of three different numbers from 1 to 6 as follows:

$$6 + 2 + 1; \quad 5 + 3 + 1; \quad 4 + 3 + 2.$$

These sums can then only go on one of the three sides of the triangle. The corner (vertex) numbers are the ones that are repeated (1, 2, 3). These sums lead to a unique answer.

Show how to do this for the other side sums.

8. (i) Take any six numbers (not necessarily whole numbers) and use them instead of $\{1, 2, 3, 4, 5, 6\}$ in the original problem; (ii) put three circles along the edge of any regular polygon; (iii) what else did you get?

9. (i) Put four circles on each side of an equilateral triangle to give another eight circle problem; (ii) try putting circles along the edge of a pyramid; (iii) put nine circles in a 3 by 3 array. Is it possible the three circled numbers in each of the rows and columns add up to the same value?

10. A set of numbers is nice, if and only it is of the form $\{a, b, c, a + d, b + d, c + d\}$.

First of all it is easy to see that if we put the numbers a, b and c in the corners of the triangle and $a + d, b + d, c + d$ in the middle of the sides opposite a, b and c, respectively, we get a snark. So $\{a, b, c, a + d, b + d, c + d\}$ is nice.

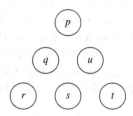

Second let the numbers p, q, r, s, t, u be put into the circles as shown above. Assume that this set of six numbers form a snark. Since

$$p + q + r = r + s + t = t + u + p$$

we can show that $p - s = r - u = t - q$, and so $\{p, q, r, s, t, u\}$ is of the form $\{a, b, c, a + d, b + d, c + d\}$.

11. Much of this is personal knowledge but we did generalise from the equilateral triangle and the right angled triangle in Chapter 2.

12. Using the results of Exercise 10, we can show by a little exhaustive algebra, that if a nice set has four associated snarks, then the numbers in the nice set are in arithmetic progression. On the other hand, take a nice set that is an arithmetic progression, say $a, a + d, a + 2d, a + 3d, a + 4d, a + 5d$. Now recall the four snarks that we got using 1, 2, 3, 4, 5, 6 in Exercise 2. If we replace 1 by a, 2 by $a + d$, and so on, we get the four snarks we would like.

13. Again it is difficult for me to write out this list for you, but I expect it will be relatively long.

14. Five ways? The way described — by diagram; by guess and check; by guess and improve (where we take the previous wrong answer and instead of just randomly guessing again, use the last guess to get a better guess); and by algebra (and there may be more than one way to do this).

    But my favourite method is to ask the pigs and turkeys to all stand in a line. Then to ask them to lift their front legs (if they have them). Now since there are 15 animals we must have 2×15 legs on the ground. So there are 50–30 legs in the air. These all belong to pigs, so there are $20 \div 2 = 10$ pigs!

15. That is not easy to do in a few lines. But if you do several special cases you might guess that you need $n^2 + 2n$ moves in general and that n^2 of these are jumps and $2n$ are slides. You then have to prove both that you can do it in this many moves (you will need to establish an algorithm) and that it can't be done in fewer moves. (See Derek Holton, The frogs problem: more than you might want to know, *The Mathematical Gazette*, vol. 93, number 528, November 2009.)

16. First show that multiples of 3 are good for the second player. Look at $n = 3$. Then if the first player takes 1, the second player takes 2 and wins. A similar thing happens if the first player takes 2. If there are 6 pieces, the second player is able to get that down to 3 in two moves. For

any multiple of 3, the second player can reduce the number of pieces by 3. So the second player wins the 21 Game no matter what the first player can do.

In the n Game if n is a multiple of 3, then the answer is the same. But what if n is not a multiple of 3?

What happens if players can take 1, 2 or 3 pieces when it is their turn?

17. How about this? Call the humans H and the aliens A. The table shows how the trips are organised. The first column shows, at each stage, who is on the original moon and the last column who is on the moon the six creatures are trying to reach. Brackets in this column show who is there before the return shuttle takes off. "Shuttle there" indicates a shuttle going to the new moon and who is on board. "Shuttle back" is the column showing who comes back.

This moon	Shuttle there	Shuttle back	That moon
HHH; AAA			
HH; AA	*HA*		*(H; A)*
HH; AA		*H*	*A*
HHH; AA			*A*
HHH;	*AA*		*(AAA)*
		A	*AA*
HHH; A			
H; A	*HH*		*(HH; AA)*
		H; A	*H; A*
HH; AA			
AA	*HH*		*(HHH; A)*
		A	*HHH*
AAA			
A	*AA*		*(HHH; AA)*
		A	*HHH; A*
AA			
	AA		*(HHH; AAA)*

And have you thought of extending or generalising this problem? What is the minimum number of shuttle journeys to make this all happen?

What happens if there are four humans and four aliens? Can they all get to the new moon?

18. Once again I'll just answer the problem. By various methods (maybe even five different methods) you can find that $a = 3$ and $b = 6$.

This problem is well worth extending and generalising. It leads to some interesting results.

Chapter 4

Number Theory 2

There is a chapter dealing with Number Theory (Number Theory 1) in Chapter 4 of *First Step* that will serve as an introduction to this one. It will especially be useful to make sure that you are familiar with the congruence/modulo notation $a \equiv b \pmod{n}$.

Mainly here I deal with Euler's Theorem and Wilson's Theorem. Euler's Theorem is a generalisation of Fermat's Little Theorem (see Chapter 4 of *First Step* and Section 4.1 below). Wilson's Theorem is somehow related. These results are largely concerned with powers of a number modulo a prime. So on with a little more Number Theory.

4.1. A Problem

In the Number Theory chapter referred to above, I talked about Fermat's Little Theorem. Just to refresh your memory, here it is again.

Fermat's Little Theorem (F.L.T.). *If p is a prime and $1 \leq a < p$, then $a^p \equiv a \pmod{p}$.*

I just want to juggle that around a bit. If $a^p \equiv a \pmod{p}$ then, what this really means is, that $a^p = a + kp$ where $k \in \mathbb{Z}$, the set of integers. So

$$kp = a^p - a = a(a^{p-1} - 1).$$

But a is smaller than p, so $(a, p) = 1$. (Remember $(a, b) = c$ means that c is the highest common factor of a and b. So $(a, p) = 1$ means that a and p have no factors in common.) Since a is a factor of the right-hand side it must be a factor of kp. Now a and p have no common factors so a must be a factor of k. So $\frac{k}{a} = k'$ is an integer.

We now know that $k'p = a^{p-1} - 1$. Writing this another way gives

$$a^{p-1} \equiv 1 \pmod{p}.$$

Hence we have a corollary (a result that follows from) to F.L.T.

Corollary 1. *If p is a prime and $1 \le a < p$, then $a^{p-1} \equiv 1 \pmod{p}$.*

The problem I want to raise now is this, for what values of a and n, with n **not** a prime, is $a^{n-1} \equiv 1 \pmod{n}$. In other words, can we generalise F.L.T.? What is special about primes that they work this way and other numbers that don't? Or maybe other numbers do!

In line with our discussions in Chapter 3, one way to tackle this problem is to try a whole stack of examples and see if we can come up with a good conjecture.

Exercises

1. For which of the following values of a and n, is $a^{n-1} \equiv 1 \pmod{n}$?
   (i) $n = 4, a = 1, 2, 3$;   (ii) $n = 6, a = 1, 2, 3, 4, 5$;
   (iii) $n = 8, a = 1, 2, 3, 4, 5, 6, 7$;   (iv) $n = 10, a = 1, 2, 3, 4, 5, 6, 7, 8, 9$;
   (v) $n = 14, a \in \{i : 1 \le i \le 13\}$;   (vi) $n = 9, a = 1, 2, 3, 4, 5, 6, 7, 8$.

   I suggest you put your calculations in a table of the form:

1	2	3	4	5
1				
2		$2^3 \equiv \cdots \pmod 3$		
3			$3^4 \equiv \cdots \pmod 4$	
$\cdots$				
$\cdots$				

Here the top row gives the power to which the number in the left column is raised. Hence the 2^3 and 3^4 positions as shown. However, in the position 2^3 put the smallest value of $2^3 \pmod n$ not 2^3. Similarly for 3^4, etc.

2. If you have access to a computer, produce tables like those in Exercise 1 for a range of values of a and n. (You can probably do this on a spreadsheet.)

Well, its pretty clear from all that work that a^{n-1} ***isn't*** congruent to 1 for very many values of a and n at all. In fact, the whole thing looks pretty much of a mess. There do, however, seem to be some funny patterns that keep turning up. First it is clear that for some a and n, $a^{n-1} \equiv 0 \pmod{n}$. Then we seem to have some n for which $a^{n-1} \equiv a \pmod{n}$ for **all** a.

What's going on here?

Exercise

3. Find under what conditions on a and n, $a^{n-1} \equiv 0 \pmod{n}$? Prove these conditions are in fact the correct ones.

   (Now is the time to use what you learnt from the last chapter. If you can't guess the answer immediately, then see what a and n you found in Exercises 1 and 2 which had this property — the simpler case heuristic. Then guess/conjecture what pattern you think always holds. Next try to prove what you've guessed. If you can't find a proof maybe there is a counter-example to your conjecture somewhere. Try to find this counter-example and readjust your conjecture.)

For a start it looks as if $a^{n-1} \equiv 0 \pmod{n}$ if and only if $n = p^\alpha$ and $a = p^\beta$, where $1 \leq \beta < \alpha$ and p is a prime. The evidence for this is $a = 2, n = 4$; $a = 2, n = 8$; and $a = 3, n = 9$. But it doesn't explain why $6^7 \equiv 0 \pmod{8}$ or $6^8 \equiv 0 \pmod{9}$.

Maybe though we can keep $n = p^\alpha$, with $\alpha > 1$. Then try $\alpha = kp^\beta$ for $1 \leq \beta < \alpha$.

Conjecture 1. $a^{n-1} \equiv 0 \pmod{n}$ *if and only if* $n = p^\alpha$ *and* $a = kp^\beta$.

Proof?   If $n = p^\alpha$ then $p^{n-1} = p^\alpha p^{n-1-\alpha} \equiv 0 \pmod{n}$. So $a^{n-1} = k^{n-1}(p^\beta)^{n-1} = k^{n-1}(p^{n-1})^\beta \equiv 0 \pmod{n}$ (because $p^{n-1} \equiv 0 \pmod{n}$).

So one direction works.

Now assume that $a^{n-1} \equiv 0 \pmod{n}$. This means n divides a^{n-1}. Hence any prime factor p of n divides a^{n-1}. Now if $p \mid n$, then $p \mid a^{n-1}$, which forces $p \mid a$. So if $a^{n-1} \equiv 0 \pmod{n}$ then $n = p^\alpha$ and $a = kp^\beta$. □

Exercises

4. The last sentence of the "proof" should start to worry you. Can you find a specific n which is not a prime power, and an a for which $a^{n-1} \equiv 0 \pmod{n}$? Can you find a general n?

5. Now state and prove a conjecture which will cover all cases for which $a^{n-1} \equiv 0 \pmod{n}$.

But you might remember that I was really trying to find out when $a^{n-1} \equiv 1 \pmod{n}$.

4.2. Euler's ϕ-function

Let's take a break from F.L.T. and learn something about Euler and his ϕ-function.

For a start, Euler was the 18th Century Swiss mathematician who we first came across in Chapter 3 of *First Step*. Among other things, he set out to solve the great Königsberg bridge mystery, you may remember. You can find more about him at www-groups.dcs.st-and.ac.uk/~history/Biographies/Euler.html.

Euler made great contributions to many areas of mathematics, not the least being to number theory. Here I want to look at an important function he introduced.

Let $\phi(n)$ be the number of positive integers less than or equal to n which are relatively prime to n. This is the **Euler ϕ-function**. (Two numbers are relatively prime if they have no common factors, except 1.)

It takes a while to get to grips with $\phi(n)$ so let's have a look at some examples. This is our experimenting phase. It should help you to get a better impression of what the ϕ-function is all about. This will help lead us to find a rule for $\phi(n)$.

Example 1.

(i) $\phi(6) = 2$. Now $(1,6) = 1, (2,6) = 2 \neq 1, (3,6) = 3 \neq 1, (4,6) = 2 \neq 1,$ $(5,6) = 1, (6,6) = 6 \neq 1$. So only 1 and 5 are less than or equal to 6 and relatively prime to 6. Hence $\phi(6) = 2$.

(ii) $\phi(9) = 6$. Now $(1,9) = 1, (2,9) = 1, (3,9) \neq 1, (4,9) = 1, (5,9) = 1,$ $(6,9) \neq 1, (7,9) = 1, (8,9) = 1$ and $(9,9) \neq 1$. So $\phi(9) = 6$ because $1, 2, 4, 5, 7$ and 8 are all relatively prime to 9 and less than or equal to 9.

Exercises

6. Find $\phi(n)$ for $1 \leq n \leq 25$.
7. If p is any prime what is $\phi(p)$?
8. Use Exercise 6 to show that $\phi(p^2) = p\phi(p)$ for $p = 2, 3, 5$. Does this formula hold for any other prime numbers p? Does $\phi(n^2) = n\phi(n)$ for any values of n apart from prime numbers?
9. If p is a prime, find an expression for $\phi(p^3)$ in terms of p.

At this stage it might be worth looking at another example.

Example 2. Show that $\phi(39) = \phi(3)\phi(13)$.

We know that $\phi(3) = 2$ and $\phi(13) = 12$. The following 24 numbers are relatively prime to 39: $1, 2, 4, 5, 7, 8, 10, 11, 14, 16, 17, 19, 20, 22, 23, 25,$ $28, 29, 31, 32, 34, 35, 37, 38$.

But let's look at those 24 numbers to see if we can see **why** there are 2×12 of them and to see **how** they are connected to $\phi(3)$ and $\phi(13)$.

First of all, any number can be written in the form $3m + r$, where $0 \leq r \leq 2$ (see Chapter 4 of *First Step*). Now $3m + r$ is relatively prime to 3 if and only if r is relatively prime to 3. This happens when $r = 1$ or 2.

What are the numbers less than 39 which are relatively prime to 3? They are the numbers $\{3m + 1 \colon 0 \leq m \leq 12\} \cup \{3m + 2 \colon 0 \leq m \leq 12\}$.

The numbers of the form $3m + 1$ are $1, 4, 7, 10, 13, 16, 19, 22, 25,$ $28, 31, 34, 37$. Of these only 13 is not relatively prime to 13. This gives 12 numbers less than 39 and relatively prime to 39. Note that $12 = \phi(13)$.

Similarly, in the set $\{3m + 2 \colon 0 \leq m \leq 12\}$, only 26 is not relatively prime to 13 and therefore to 39. Hence we have another $12 = \phi(13)$ numbers which are relatively prime to 39.

Hence $\phi(39) = 24 = 2 \times 12 = \phi(3)\phi(13)$. Note that there are $\phi(3)$ sets $\{3m + 1\}$, $\{3m + 2\}$ relatively prime to 3 and in each of these sets there are $\phi(13)$ numbers relatively prime to 13.

Exercises

10. Use the arguments of Example 1 to show that
    (i) $\phi(15) = \phi(3)\phi(5)$;   (ii) $\phi(20) = \phi(4)\phi(5)$;
    (iii) $\phi(60) = \phi(4)\phi(15)$;   (iv) $\phi(210) = \phi(14)\phi(15)$.
11. Can the arguments of Example 1 be used to show that $\phi(24) = \phi(4)\phi(6)$?
12. (a) Prove that if p and q are primes with $q = ps + 1$ for some s, then
       $\phi(pq) = \phi(p)\phi(q)$.
    (b) Prove that if p and q are primes, then $\phi(pq) = \phi(p)\phi(q)$.
13. (a) For what $m, n \leq 25$ is it true that $\phi(mn) = \phi(m)\phi(n)$?
    (b) In general, for what m, n is it true that $\phi(mn) = \phi(m)\phi(n)$?
14. (a) If p, q, r are prime numbers, find an expression for $\phi(pqr)$ in terms of
       p, q, r.
    (b) Repeat (a) with (i) $\phi(p^2 q)$; (ii) $\phi(p^2 q^3 r)$; (iii) $\phi(p^\alpha)$.

So can we find a formula for $\phi(n)$, where $n = p_1^{\alpha_1} p_2^{\alpha_2} \cdots p_r^{\alpha_r}$?

Theorem 1. *If* $n = p_1^{\alpha_1} p_2^{\alpha_2} \cdots p_r^{\alpha_r}$, *where the* p_i *are distinct primes and the* α_i *are natural numbers then* $\phi(n) = n(1 - \frac{1}{p_1})(1 - \frac{1}{p_2}) \cdots (1 - \frac{1}{p_r})$.

Proof.   Since $p_1^{\alpha_1}$ is relatively prime to $\frac{n}{p_1^{\alpha_1}}$, then by Exercise 13,

$\phi(n) = \phi(p_1^{\alpha_1})\phi(p_2^{\alpha_2} \cdots p_r^{\alpha_r})$. But the same argument allows us to write

$\phi(n) = \phi(p_1^{\alpha_1})\phi(p_2^{\alpha_2})\phi(p_3^{\alpha_3} \cdots p_r^{\alpha_r})$. Repeating this argument again

and again gives

$\phi(n) = \phi(p_1^{\alpha_1})\phi(p_2^{\alpha_2}) \cdots \phi(p_r^{\alpha_r})$.

But from Exercise 14 we know that $\phi(p^\alpha) = p^\alpha - p^{\alpha-1}$, for p a prime. Hence

$$\phi(n) = \phi(p_1^{\alpha_1})\phi(p_2^{\alpha_2}) \cdots \phi(p_r^{\alpha_r})$$
$$= (p_1^{\alpha_1} - p_1^{\alpha_1-1})(p_2^{\alpha_2} - p_2^{\alpha_2-1}) \cdots (p_r^{\alpha_r} - p_r^{\alpha_r-1})$$
$$= p_1^{\alpha_1}\left(1 - \frac{1}{p_1}\right)p_2^{\alpha_2}\left(1 - \frac{1}{p_2}\right) \cdots p_r^{\alpha_r}\left(1 - \frac{1}{p_r}\right)$$
$$= p_1^{\alpha_1}p_2^{\alpha_2} \cdots p_r^{\alpha_r}\left(1 - \frac{1}{p_1}\right)\left(1 - \frac{1}{p_2}\right) \cdots \left(1 - \frac{1}{p_r}\right)$$
$$= n\left(1 - \frac{1}{p_1}\right)\left(1 - \frac{1}{p_2}\right) \cdots \left(1 - \frac{1}{p_r}\right). \qquad \square$$

Exercises

15. Armed with Theorem 1, go back and solve the questions raised in Exercises 8 and 9.

16. (a) For what values of $2n \leq 25$ is $\phi(2n) = \phi(n)$?
    (b) For what n is $\phi(2n) = \phi(n)$?
    (c) For what n is $\phi(2n) = n$?

17. Go back to the answers to Exercise 6. Look at the range of possible values for $\phi(n)$.
    (a) Find $\{n : \phi(n) = 2\}$.
    (b) Find $\{n : \phi(n) = 4\}$.
    (c) Find $\{n : \phi(n) = 12\}$.
    (d) For $n > 2$ what do you conjecture about the possible values of $\phi(n)$?
    (e) Is $\{n : \phi(n) = 2m$ for m fixed$\}$
        (i) non-empty for all m;   (ii) finite for all m?

4.3. Back to Section 4.1

You may have forgotten but several pages and several problems ago I wanted to see if Fermat's Little Theorem ($a^{p-1} \equiv 1 \pmod{p}$) could be generalised to numbers which were not prime. For what values a and n with n **not** prime is $a^{n-1} \equiv 1 \pmod{n}$? We started to find evidence on this problem way back in Exercise 1. Go back and see what results you got there.

Have a look at $n = 6$. The only non-trivial value of a worth looking at is $a = 5$. But $5^5 \not\equiv 1 \pmod 6$. However 5^2 is congruent to 1 (mod 6) as is 5^4. Is there anything significant about that power of 2 (or 4)?

Exercises

18. (a) For what a is $a^7 \equiv 1 \pmod 8$?
    For what a and what m is $a^m \equiv 1 \pmod 8$?
    Are there any significant numbers among the m?
    (b) Repeat (a) with 8 replaced by 10.
    (c) Repeat (a) with 8 replaced by 14.
    (d) Repeat (a) with 8 replaced by 9.
19. In the light of all you have read in this chapter so far, produce a conjecture for what values of a and m force $a^m \equiv 1 \pmod n$.

First of all it should by now be obvious that the only way for a^m to be congruent to 1 modulo n is for (a, n) to be 1. Suppose this wasn't the case. Then let $(a, n) = r > 1$, where $a = \alpha r$ and $n = \mu r$.

If $a^m \equiv 1 \pmod n$, then $\alpha^m r^m = q \mu r + 1$. Hence $1 = (\alpha^m r^{m-1} - q\mu)r$. But this plainly cannot be. The right hand side is clearly divisible by r while the left hand side isn't.

So from now on we'll restrict our attention to those a for which $(a, n) = 1$.

What about m though? What can it be? (Not surprisingly more questions — always more questions. It's the only way to get ahead. And notice how we again creep up on a general result by looking at a host of special cases.)

Exercises

20. (a) Show that $a^4 \equiv 1 \pmod{12}$ for all a such that $1 \le a \le 11$ and $(a, 12) = 1$.
    (b) Show that $a^8 \equiv 1 \pmod{15}$ for all a such that $1 \le a \le 14$ and $(a, 15) = 1$.
    (c) Show that $a^6 \equiv 1 \pmod{14}$ for all a such that $(a, 14) = 1$.
21. Solve the following equations, where x is always between 1 and n, the number of the modulus.
    (i) $3x \equiv 1 \pmod 4$; (ii) $2x \equiv 3 \pmod 5$;
    (iii) $4x \equiv 5 \pmod 7$; (iv) $5x \equiv 2 \pmod 9$.

22. (a) For $(x, n) = 1$, show that if $tx \equiv x$ (mod n), then $t \equiv 1$ (mod n).
    (b) Show that if $(a, n) = 1$ and $ay \equiv 0$ (mod n), then $y \equiv 0$ (mod n).
    (c) Show that if $(a, n) = 1$ and $(b, n) = 1$ then $(ab, n) = 1$.
23. (a) Let $T = \{x : 1 \leq x \leq 7 \text{ and } (x, 8) = 1\}$. Let $t \in T$. Show that for every $x \in T$, there is an $x' \in T$ such that $tx \equiv x'$ (mod 8).
    (b) Use (a) to show that $t^4 \equiv 1$ (mod 8) for all $t \in T$.
24. (a) Let $T = \{x : 1 \leq x \leq 8 \text{ and } (x, 9) = 1\}$. Let $t \in T$. Show that $\{x' : tx \equiv x'$ (mod 9) for $x \in T\} = T$.
    (b) Use (a) to show that $t^6 \equiv 1$ (mod 9).
25. Show that for all a such that $(a, n) = 1$, $a^{\phi(n)} \equiv 1$ (mod n).

Euler spent a lot of time thinking mathematical thoughts (as well as walking over bridges). Much of this time was spent on Number Theory. We have him to blame for $a^{\phi(n)} \equiv 1$ (mod n). The proof of Euler's Theorem is a little long, but try to follow it. Come back again later and have another go at it.

Theorem (Euler). *If a is such that $(a, n) = 1$, then $a^{\phi(n)} \equiv 1$ (mod n).*

Proof. Let $t = \{x_1, x_2, \ldots, x_{\phi(n)}\}$ be the set of numbers from 1 to $n - 1$ inclusive which are relatively prime to n. (Clearly there are $\phi(n)$ of them.)

Let a be any number such that $(a, n) = 1$. If $ax_i \equiv a_i$ (mod n) for any $x_i \in T$ then what or where is a_i? Let $T' = \{a_i : ax_i \equiv a_i$ (mod n) with $x_i \in T\}$.

Step 1. $T' = T$:
First of all it's obvious that if $ax_i = ax_j$, then $x_i = x_j$.

Further if $ax_i \equiv a_i$ (mod n) and $ax_j \equiv a_j$ (mod n), does $a_i = a_j$ imply $x_i = x_j$? Well, if $a_i = a_j$, then $ax_i = ax_j + kn$ for some k. Hence $a(x_i - x_j) = kn$. By Exercise 22(b) this implies that $x_i - x_j \equiv 0$ (mod n). Since $|x_i - x_j| < n$, then $x_i = x_j$. So $a_i = a_j$ does imply $x_i = x_j$.

What this means then, is that $|T'| = |T|$. This is because we get a different a_i for every $x_i \in T$.

But $(a, n) = 1$ and $(x_i, n) = 1$. Hence by Exercise 22(c), $(ax_i, n) = 1$. So, clearly, $(a_i, n) = 1$.

This means that T' contains numbers which are relatively prime to n. But so does T, in fact it contains all of them. So $T' \subseteq T$. However, $|T'| = |T|$. Hence $T' = T$.

Step 2. $a^{\phi(n)} \equiv 1$ (mod n):
Since $T' = T$ it means that

$$(ax_1)(ax_2) \cdots (ax_{\phi(n)}) \equiv a_1 a_2 \cdots a_{\phi(n)} \equiv x_1 x_2 \cdots x_{\phi(n)} (\text{mod } n).$$

So $a^{\phi(n)}x_1x_2\cdots x_{\phi(n)} \equiv x_1x_2\cdots x_{\phi(n)}$ (mod n). If we let $x_1x_2\cdots x_{\phi(n)} = x$, then $a^{\phi(n)}x \equiv x$ (mod n). But $x \equiv 1$ (mod n) by repeated use of Exercise 22(c). Hence $a^{\phi(n)} \equiv 1$ (mod n) by Exercise 22(a). □

Exercises

26. Is $\phi(n)$ the smallest value of m for which $a^m \equiv 1$ (mod n)?

27. Show that Fermat's Little Theorem follows from Euler's Theorem.

28. Find the least positive integer b which satisfies the following congruences:
    (i) $3^{56} \equiv b$ (mod 7);   (ii) $3^{56} \equiv b$ (mod 11);
    (iii) $7^{32} \equiv b$ (mod 7);   (iv) $7^{32} \equiv b$ (mod 11);
    (v) $7^{122} \equiv b$ (mod 11);   (vi) $7^{122} \equiv b$ (mod 13).

29. If $(a, 2p) = 1$, where p is a prime, what can be said about a^{2p-1}? Can this be generalised? (There is a lot in this one. You may need to spend some time on it.)

30. (a) Show that $a^{13} - a \equiv 0$ (mod 2730) for every integer a.
    (b) Show that $a^{17} - a \equiv 0$ (mod 8160) for every odd integer a.
    Both of these require quite a bit of work.

4.4. Wilson

At one stage in the proof of Euler's Theorem we used $x_1x_2\cdots x_{\phi(n)}$ but didn't bother to find out exactly what value it had.

Let's investigate that product now.

Exercises

31. For arbitrary n, let $x_1x_2\cdots x_{\phi(n)} \equiv r_n$ (mod n), where $(x_i, n) = 1$ for all x_i. Find r_n for
    (i) $n = 3$;   (ii) $n = 4$;   (iii) $n = 5$;
    (iv) $n = 6$;   (v) $n = 7$;   (vi) $n = 8$;
    (vii) $n = 9$;   (viii) $n = 10$;   (ix) $n = 11$.

32. Conjecture the value of r_p, where p is a prime. Check out your conjecture on primes which don't occur in Exercise 31. Prove your conjecture.

33. Conjecture the value of r_n, where n is not a prime. Test your conjecture on various values of n.
    Prove your conjecture.

Let's have a look at $p = 67$. The product $x_1x_2\cdots x_{\phi(67)} = 1\cdot2\cdot3\cdots\cdots 66 = 66!$. It will take a bit of computing to work out what 66! is congruent to modulo 67. Is there a quicker way?

Obviously $(2, 67) = 1$. So by Theorem 1 on page 6 of Chapter 1 of *First Step*, there is some a and b such that $2a + 67b = 1$. So $2a \equiv 1 \pmod{67}$. There must, in fact, be some $a \leq 67$ (and positive) for which $2a \equiv 1 \pmod{67}$.

Now $a = 34$ is an obvious solution to the last equation. Can there be any other? If $2a \equiv 1 \pmod{67}$ and $2a' \equiv 1 \pmod{67}$, then $2a \equiv 2a' \pmod{67}$. Multiplying both sides by 34 gives $a \equiv a' \pmod{67}$. So the only solution of $2a \equiv 1 \pmod{67}$ for which $1 \leq a \leq 67$ is $a = 34$.

Now have a look at 3. Again $(3, 67) = 1$ and again $3a + 67b = 1$ for some a and b. Hence $3a \equiv 1 \pmod{67}$. Here $a = 45$ and this a is unique.

Is it true that for any r with $(r, 67) = 1$ that there is a unique a_r such that $ra_r \equiv 1 \pmod{67}$? If there is, then we can pair up each r with its a_r in the product 66! and know that their product is congruent to 1.

For instance $66! = 1 \cdot 2 \cdot 3 \cdots 66$
$$= 1 \cdot (2 \cdot 34)(3 \cdot 45)(4 \cdot 17) \cdots (\ ).$$

Wait on! There are an even number of numbers between 1 and 66 inclusive. Now if 1 stands by itself (since only $1 \cdot 1 \equiv 1 \pmod{67}$), then not **all** of the other numbers can pair themselves into products congruent to 1.

Which number or numbers don't have pairs?

Exercises

34. Show that 66 and 1 are the only numbers r between 1 and 66 which have no other number a_r with the property that $ra_r \equiv 1 \pmod{67}$.
35. Show that 1 and 52 are the only numbers r $(1 \leq r \leq 52)$ which have no other number a_r with the property that $ra_r \equiv 1 \pmod{53}$.
36. For any odd prime p, show that any r $(2 \leq r \leq p - 2)$ has an a_r $(2 \leq a_r \leq p - 2)$ with $r \neq a_r$, such that $ra_r \equiv 1 \pmod{p}$

In view of Exercise 34,

$$66! = 1 \cdot 2 \cdot 3 \cdots 66$$
$$= 1 \cdot (2 \cdot 34)(3 \cdot 45)(4 \cdot 17) \cdots (32 \cdot 44)66$$
$$\equiv 1 \cdot 66 \pmod{67}$$
$$\equiv 66 \pmod{67}$$
$$\equiv -1 \pmod{67}.$$

In fact, if we are to believe Exercise 36 we must have $(p - 1)! \equiv -1 \pmod{p}$ for **any** odd prime p. The argument is exactly the same as for 67.

In $(p-1)!$ simply pair up numbers r and a_r for which $ra_r \equiv 1 \pmod{p}$. The only two numbers which don't pair in such a way are 1 and $p-1$. Hence

$$(p-1)! = 1(2 \cdot a_2)(3 \cdot a_3) \cdots (\ )(p-1)$$
$$\equiv p-1 \pmod{p}$$
$$\equiv -1 \pmod{p}.$$

We have succeeded in proving Wilson's Theorem.

Theorem (Wilson). *If p is an odd prime, $(p-1)! \equiv -1 \pmod{p}$.*

Exercises

37. What is $(p-1)!$ congruent to, modulo p, for p an even prime?
38. (a) If p is a prime such that $p \equiv 1 \pmod 4$, show that $[(\frac{p-1}{2})!]^2 \equiv -1 \pmod p$.
    (b) If p is a prime such that $p \equiv 3 \pmod 4$, show that $[(\frac{p-1}{2})!]^2 \equiv 1 \pmod p$.
39. Show that $(p-1)! \equiv -1 \pmod p$ if and only if p is an odd prime.

4.5. Some More Problems

Having worked hard to establish Euler's Theorem and Wilson's Theorem here's your chance to use these results. Below is a selection, in no particular order, of some harder problems from IMO's and other sources. One of these "other sources" is the journal "*Crux Mathematicorum*". This is a publication of the Canadian Mathematical Society. (See http://journals.cms.math.ca/CRUX/.)

Exercises

40. Let $A = a^4$ where a is a positive integer. Find all positive integers x such that $A^{15x+1} \equiv A \pmod{6814407600}$.
    (Crux Math. [1987:119].)
41. Let p be a prime number such that $p > 3$. Show that p^3 divides $\binom{2p}{p} - 2$.
    (This is an extension of the 1987 Australian Mathematical Olympiad Question 2.)
42. (a) Find all positive integers n such that $2^n - 1$ is divisible by 7.

(b) Prove that there is no positive integer n for which $2^n + 1$ is divisible by 7.

(IMO 1964 Question 1.)

43. Let m be the smallest positive integer for which $a^m \equiv 1 \pmod{n}$. Show that if $a^t \equiv 1 \pmod{n}$ then $m \mid t$.

44. m and n are natural numbers with $1 \le m < n$. In their decimal representations, the last three digits of 1978^m are equal, respectively, to the last three digits of 1978^n. Find m and n such that $m + n$ has its least value.

(IMO 1978 Question 1.)

45. Find one pair of positive integers a and b such that
   (i) $ab(a + b)$ is not divisible by 7;
   (ii) $(a + b)^7 - a^7 - b^7$ is divisible by 7^3.

(IMO 1984 Question 2.)

46. Show that, for p and odd prime,

$$p^3 \left| \binom{2p}{p} - 2. \right.^{\text{a}}$$

4.6. Solutions

1. The numbers a for which $a^{n-1} \equiv 1 \pmod{n}$ are underlined.

(i) $n = 4$

1	2	3
<u>1</u>	<u>1</u>	<u>1</u>
2	0	0
3	1	3

(ii) $n = 6$

1	2	3	4	5
<u>1</u>	1	1	1	1
2	4	2	4	2
3	3	3	3	3
4	4	4	4	4
<u>5</u>	1	5	1	5

(iii) $n = 8$

1	2	3	4	5	6	7
<u>1</u>	1	1	1	1	1	1
2	4	0	0	0	0	0
<u>3</u>	1	3	1	3	1	3
4	0	0	0	0	0	0
<u>5</u>	1	5	1	5	1	5
6	4	0	0	0	0	0
<u>7</u>	1	7	1	7	1	7

---

[a] $\binom{n}{r} = {}^nC_r$, see page 41 of *First Step*.

(iv) $n = 10$

	1	2	3	4	5	6	7	8	9
1	1	1	1	1	1	1	1	1	1
2	2	4	8	6	2	4	8	6	2
3	3	9	7	1	3	9	7	1	3
4	4	6	4	6	4	6	4	6	4
5	5	5	5	5	5	5	5	5	5
6	6	6	6	6	6	6	6	6	6
7	7	9	3	1	7	9	3	1	7
8	8	4	2	6	8	4	2	6	8
9	9	1	9	1	9	1	9	1	9

(v) $n = 14$

	1	2	3	4	5	6	7	8	9	10	11	12	13
1	1	1	1	1	1	1	1	1	1	1	1	1	1
2	2	4	8	2	4	8	2	4	8	2	4	8	2
3	3	9	13	11	5	1	3	9	13	11	5	1	3
4	4	2	8	4	2	8	4	2	8	4	2	8	4
5	5	11	13	9	3	1	5	11	13	9	3	1	5
6	6	8	6	8	6	8	6	8	6	8	6	8	6
7	7	7	7	7	7	7	7	7	7	7	7	7	7
8	8	8	8	8	8	8	8	8	8	8	8	8	8
9	9	11	1	9	11	1	9	11	1	9	11	1	9
10	10	2	6	4	12	8	10	2	6	4	12	8	10
11	11	9	1	11	9	1	11	9	1	11	9	1	11
12	12	4	6	2	10	8	12	4	6	2	10	8	12
13	13	1	13	1	13	1	13	1	13	1	13	1	13

(vi) $n = 9$

	1	2	3	4	5	6	7	8
1	1	1	1	1	1	1	1	1
2	2	4	8	7	5	1	2	4
3	3	0	0	0	0	0	0	0
4	4	7	1	4	7	1	4	7
5	5	7	8	4	2	1	5	7
6	6	0	0	0	0	0	0	0
7	7	4	1	7	4	1	7	4
8	8	1	8	1	8	1	8	1

There are some interesting patterns here both for what we are looking for and for other things.

2. This is between you and your computer.

3. Is it true that $a^{n-1} \equiv 0 \pmod{n}$ if and only if a is a factor of n? Do we have to be more restrictive than this?

4. Try $n = 12$ (not a prime power) and $a = 6$ for the first part. Now $6^2 \equiv 0 \pmod{12}$, so $6^{11} \equiv 0 \pmod{12}$. This wrecks Conjecture 1. So where did my "proof" go wrong? You need to check my every move!

What general n do you suggest? How about $n = p_1^{\alpha_1} p_2^{\alpha_2} \cdots p_r^{\alpha_r}$, where the p_i are primes and the α_i are natural numbers (so none of them are zero). For such n can we always find an a for which $a^{n-1} \equiv 0 \pmod{n}$?

5. **Conjecture 2.** $a^{n-1} \equiv 0 \pmod{n}$ *if and only if* $n = p_1^{\alpha_1} p_2^{\alpha_2} \cdots p_r^{\alpha_r}$ *where the p_i are primes and the α_i are natural numbers and* $a = p_1^{\beta_1} p_2^{\beta_2} \cdots p_r^{\beta_r}$, *where* $1 \le \beta_i \le \alpha_i$.

There's actually still one flaw in this conjecture. Find it and prove your adjusted conjecture.

(The correct result is proved later.)

6.

n	1	2	3	4	5	6	7	8	9	10
$\phi(n)$	1	1	2	2	4	2	6	4	6	4
n	11	12	13	14	15	16	17	18	19	20
$\phi(n)$	10	4	12	6	8	8	16	6	18	8
n	21	22	23	24	25					
$\phi(n)$	12	10	22	8	20					

7. $\phi(p) = p - 1$. Every integer between 1 and $p - 1$ is relatively prime to p.

8. From Exercise 6, $\phi(4) = 2 = 2\phi(2); \phi(9) = 6 = 3\phi(3); \phi(25) = 20 = 5\phi(5)$.

If p is a prime $\phi(p^2) = p\phi(p)$. To see this, consider what numbers between 1 and p^2 have a factor in common with p^2. Since p is prime, these can only be $p, 2p, 3p, \ldots, (p-1)p, p^2$. Hence $\phi(p^2) = p^2 - p = p(p-1)$. From Exercise 7, since p is prime, $\phi(p) = p - 1$. Hence if p is prime, $\phi(p^2) = p\phi(p)$. (So $\phi(p^2) = p(p-1)$.)

$$\phi(4^2) = 4\phi(4) : \phi(36) = 6\phi(6).$$

Is it true that $\phi(n^2) = n\phi(n)$ for **all** n? (I'll come back to this later. In the meantime see whether or not you can prove it.)

9. Again look at the problem back-to-front. How many numbers from 1 to p^3 have factors *in common* with p^3? Surely they are just $p, 2p, 3p, \ldots, (p^2 - 1)p, p^3$. There are p^2 of these numbers. Hence $\phi(p^3) = p^3 - p^2$.

(If you like, $\phi(p^3) = p^2\phi(p)$. Is $\phi(p^\alpha) = p^{\alpha-1}\phi(p)$? Is $\phi(n^3) = n^2\phi(n)$ for all n? Is $\phi(n^\alpha) = n^{\alpha-1}\phi(n)$ for all n?)

10. (i) The numbers less than 15 which are relatively prime to 3 are in the set $\{3m + 1 : 0 \le m \le 4\}$ and $\{3m + 2 : 0 \le m \le 4\}$. In the first set these numbers are $1, 4, 7, 10, 13$. Only 10 is **not** relatively prime to 5. This leaves four numbers $(4 = \phi(5))$ which are relatively prime to 5.

In the second set we have $2, 5, 8, 11, 14$ and only 5 is not relatively prime to 5. This leaves four numbers $(4 = \phi(5))$ relatively prime to 5.

Hence $\phi(15) = 8 = \phi(3)\phi(5)$.

(ii) The numbers relatively prime to 4 are $\{4m + 1 : 0 \le m \le 4\}$ and $\{4m + 3 : 0 \le m \le 4\}$. In the first set only 5 is not relatively prime to 5 and in the second set only 15 is not relatively prime to 5. Hence $\phi(20) = 2 \times 4 = \phi(4)\phi(5)$.

(iii) Consider the sets

$$\{4m + 1 : 0 \le m \le 14\} \text{ and } \{4m + 3 : 0 \le m \le 14\}.$$

From the first set we need to delete 5, 9, 21, 25, 33, 45 and 57 to leave 8 numbers relatively prime to 60. Note that $8 = \phi(15)$.

Similarly we need to delete 3, 15, 27, 35, 39, 51 and 55 to leave $8 = \phi(15)$ numbers which are relatively prime to 60.

Hence $\phi(60) = 2 \times 8 = \phi(4)\phi(15)$.

(iv) $\{14m + 1\} \cup \{14m + 3\} \cup \{14m + 5\} \cup \{14m + 9\} \cup \{14m + 11\} \cup \{14m + 13\}$ where $0 \le m \le 14$ are the $\phi(14)$ sets which contain only numbers less than 210 which are relatively prime to 14.

From each set, when we remove the numbers not relatively prime to 15, we are left with $\phi(15)$ numbers. Hence $\phi(210) = \phi(14)\phi(15)$.

11. $\{4m + 1 : 0 \le m \le 5\} \cup \{4m + 3 : 0 \le m \le 5\}$ are all the numbers less than 24 which are relatively prime to 4.

Removing the numbers which are *not* relatively prime to 6 from $\{4m + 1\}$ leaves 1, 5, 13, 17. Hence there are four numbers in $\{4m + 1\}$ which *are* relatively prime to 24 and $4 \ne \phi(6)$.

A similar thing happens with $\{4m + 3\}$.

12. (a) Suppose $q = ps + 1$ for some integer s.

We first repeat the argument of Example 1. All the numbers less than pq which are relatively prime to p are in one of the $\phi(p)$ sets $\{pm+1 : 0 \le m < q-1\}, \{pm+2 : 0 \le m \le q-1\}, \ldots, \{pm+(p-1) : 0 \le m \le q - 1\}$.

Now since $q = ps + 1, q \in \{pm + 1 : 0 \le m \le q - 1\}$. However, no other multiple of q less than pq, lies in this set.

So $\{pm+1 : 0 \le m \le q-1\}\backslash\{q\}$ contains $q-1$ numbers relatively prime to pq. Note that $q - 1 = \phi(q)$.

Similarly since $q = ps + 1, 2q \in \{pm + 2 : 0 \le m \le q - 1\}$. Hence $\{pm + 2 : 0 \le m \le q - 2\}\backslash\{2q\}$ is a set of $q - 1 = \phi(q)$ numbers all relatively prime to pq.

The argument can be repeated on all $\phi(p)$ sets

$$\{pm + r : 0 \le m \le q - 1\}$$

for $0 \le r \le p - q$. Hence $\phi(pq) = \phi(p)\phi(q)$.

(b) What if $q \neq ps+1$? What we need to do is to show that if $q = ps+t$ for $0 \leq t \leq p-1$, that only **one** multiple of q lies in each of the sets $\{pm+r\}$. If we can do that the argument follows in the same way that it did in (a).

So assume $\alpha q = pm_1 + r$ and $\beta q = pm_2 + r$. Then $(\alpha - \beta)q = p(m_1 - m_2)$. Since $(p, q) = 1$, then $p|(\alpha - \beta)$. But $\alpha < p$ and $\beta < p$ since αq and βq are both less than pq. Hence the only way that p can divide $\alpha - \beta$ is if $\alpha - \beta = 0$. So $\alpha q = \beta q$ and there is only one multiple of q in the set $\{pm + r : 0 \leq m \leq q - 1\}$.

What if **no** multiple of q lies in $\{pm + r : 0 \leq m \leq q - 1\}$? Then the pigeonhole principle says that two multiples of q must lie in one of the other sets $\{pm + r' : 0 \leq m \leq q - 1\}$. This contradicts the last paragraph. Hence one multiple of q lies in each of the sets $\{pm + r\}$.

13. (a) $\phi(6) = \phi(2)\phi(3)$: $\phi(10) = \phi(2)\phi(5)$: $\phi(12) = \phi(3)\phi(4)$ but
    $\phi(12) \neq \phi(2)\phi(6)$: $\phi(14) = \phi(2)\phi(7)$: $\phi(18) = \phi(2)\phi(9)$ but
    $\phi(18) \neq \phi(3)\phi(6)$: $\phi(20) = \phi(4)\phi(5)$ but $\phi(20) \neq \phi(2)\phi(10)$:
    $\phi(21) = \phi(3)\phi(7)$: $\phi(22) = \phi(2)\phi(11)$: $\phi(24) = \phi(3)\phi(8)$ but
    $\phi(24) \neq \phi(2)\phi(12)$ or $\phi(4)\phi(6)$.

    (b) $\phi(mn) = \phi(m)\phi(n)$ if and only if $(m, n) = 1$. The proof of this follows the lines of the proof in Exercise 12(b). The details can be found in "classical" books on Number Theory such as the one by Hardy and Wright (*An Introduction to the Theory of Numbers*, Oxford University Press, 1979).

14. (a) $\phi(pqr) = \phi(p)\phi(qr)$ since $(p, qr) = 1$. Similarly $\phi(qr) = \phi(q)\phi(r)$. Hence $\phi(pqr) = \phi(p)\phi(q)\phi(r)$. Since p, q, r are prime, $\phi(p) = p - 1$, $\phi(q) = q - 1$, $\phi(r) = r - 1$ (see Exercise 7).
    Hence $\phi(pqr) = (p - 1)(q - 1)(r - 1)$.

    (b) (i) $\phi(p^2q) = p(p - 1)(q - 1)$;
        (ii) $\phi(p^2q^3r) = p(p - 1)q^2(q - 1)(r - 1)$;
        (iii) The arguments of Exercise 9, give $\phi(p^\alpha) = p^{\alpha-1}(p - 1)$.

15. In Exercise 8, I asked if $\phi(n^2) = n\phi(n)$. Now if $n = p_1^{\alpha_1} \cdots p_r^{\alpha_r}$, $\phi(n^2) = n^2(1 - \frac{1}{p_1})(1 - \frac{1}{p_2}) \cdots (1 - \frac{1}{p_r})$ by Theorem 1. This is clearly $n\phi(n)$ as requested.

    In the solution to Exercise 9, I asked if $\phi(n^\alpha) = n^{\alpha-1}\phi(n)$. Now

$$\phi(n^\alpha) = n^\alpha \left(1 - \frac{1}{p_1}\right) \cdots \left(1 - \frac{1}{p_r}\right)$$

$$= n^{\alpha-1}\left[n\left(1 - \frac{1}{p_1}\right) \cdots \left(1 - \frac{1}{p_r}\right)\right] = n^{\alpha-1}\phi(n).$$

16. (a) $n = 1, 3, 5, 7, 9, 11$.

   (b) From (a) we could conjecture that $\phi(2n) = \phi(n)$ if and only if n is odd. Suppose n is odd. Then $(2, n) = 1$. So $\phi(2n) = \phi(2)\phi(n) = \phi(n)$.

   Now suppose $\phi(2n) = \phi(n)$ and n is even. Then let $n = 2^{\alpha_1} p_2^{\alpha_2} \cdots p_r^{\alpha_r}$. Hence $2n = 2^{\alpha_1+1} p_2^{\alpha_2} \cdots p_r^{\alpha_r}$. Then $\phi(2n) = 2n(1 - \frac{1}{2})(1 - \frac{1}{p_1}) \cdots (1 - \frac{1}{p_r}) = 2\phi(n)$. This contradicts the fact that $\phi(2n) = \phi(n)$ so n must be odd.

   (Actually I've just proved that if n is odd $\phi(2n) = \phi(n)$, while if n is even $\phi(2n) \neq 2\phi(n)$.)

   (c) Let $2n = 2^{\alpha_1} p_2^{\alpha_2} \cdots p_r^{\alpha_r}$. We therefore require

   $$2^{\alpha_1-1} p_2^{\alpha_2-1}(p_2 - 1) \cdots p_r^{\alpha_r-1}(p_r - 1) = 2^{\alpha_1-1} p_2^{\alpha_2} \cdots p_r^{\alpha_r}.$$

   This is equivalent to $(p_2 - 1) \cdots (p_r - 1) = p_2 \cdots p_r$. Clearly the left side of this equation is smaller than the right side. Hence $\phi(2n) = n$ implies that n has no prime factors other than 2. So $n = 2^{\alpha_1-1}$ for some $a_1 \geq 1$.

17. (a) Let $n = p_1^{\alpha_1} \cdots p_r^{\alpha_r}$. Then

   $$2 = \phi(n) = p_1^{\alpha_1-1}(p_1 - 1)p_2^{\alpha_2-1}(p_2 - 1) \cdots p_r^{\alpha_r-1}(p_r - 1).$$

   Clearly n has no prime divisors greater than 3.

   Suppose $p_1 = 2$. If $\alpha_1 - 1 = 0, p_2 = 3$ and $\alpha_2 = 1$. Here $n = 6$. If $\alpha_1 - 1 = 1$, then n contains no other prime divisors. Hence $n = 4$.

   Suppose $p_1 = 3$. Then $\alpha_2 - 1 = 0$ and $n = 3$. Hence

   $$\{n : \phi(n) = 2\} = \{3, 4, 6\}.$$

   (b) Now $4 = p_1^{\alpha_1-1}(p_1-1) \cdots p_r^{\alpha_r-1}(p_r-1)$. Clearly n contains no prime divisors greater than 5.

   Suppose $p_1 = 2$. If $\alpha_1 - 1 = 2$, then $n = 8$. If $\alpha_1 - 1 = 1$, then $p_2 = 3, \alpha_2 = 1$ and n contains no other prime divisors. Hence $n = 12$. If $\alpha_1 - 1 = 0, p_2 = 5$ and $\alpha_2 = 1$. This gives $n = 10$.

   Suppose $p_1 = 3$. Then $\alpha_1 - 1 = 0$. But now the equation cannot be satisfied, no matter what p_2 equals.

   Suppose $p_1 = 5$. Then $\alpha_1 - 1 = 0$ and $n = 5$.

   Hence $\{n : \phi(n) = 4\} = \{5, 8, 10, 12\}$.

   (c) By arguments similar to those in (a) and (b) we see that

   $$\{n : \phi(n) = 12\} = \{13, 21, 26, 28, 36\}.$$

   (d) $\phi(n)$ is even. Why? Any other ideas?

   (e) The question really says, "It's easy to show that for $n \geq 3, \phi(n)$ is even. OK, so is every even number the ϕ of some n?"

(i) For what n is $\phi(n) = 14$? If n is a prime, then $\phi(n) = n - 1$. In this case this implies $n = 15$ but of course 15 is not a prime. So n must be composite. Let $n = rs$ where $(r, s) = 1$. Then $\phi(n) = \phi(r)\phi(s)$. This implies that $\phi(s)$, say, equals 7. But ϕ can only be odd if $n = 1$ or 2. So no such s exists. Hence there is no n for which $\phi(n) = 14$.

(Now find an infinite set of even numbers $2m$, such that there is no n with $\phi(n) = 2m$.)

We have just shown that $\{n : \phi(n) = 2m\}$ may be empty.

(ii) Now ask the question the other way around. "Are there any m for which $\{n : \phi(n) = 2m\}$ is **not** finite?" What a silly question. I can't even grace that question with a proof. Clearly (?) such sets must be finite. Of course a proof would be nice. Couldn't we base it on the fact that there are only a finite number of primes that would satisfy the equation $2m = p_1^{a_1-1}\cdots$?

And is it possible to say how big they can get? Can you find one that is bigger than 472 or 4,720 or any fixed number of your choice?

18. (a) No a other than 1. If $a = 3, m = 2, 4$, or 6: if $a = 5, m = 2, 4$, or 6: if $a = 7, m = 2, 4$, or 6. Undoubtedly there are significant numbers amongst the m.

(b) If $a = 3, m = 4, 8$; if $a = 7, m = 4, 8$; if $a = 9, m = 2, 4, 6, 8$.

(c) If $a = 3, m = 6, 12$; if $a = 5, m = 6, 12$; if $a = 9, m = 3, 6, 9, 12$; if $a = 11, m = 3, 6, 9, 12$; if $a = 13, m = 2, 4, 6, 8, 10, 12$.

(d) If $a = 4, m = 3, 6$; if $a = 7, m = 3, 6$; if $a = 8, m = 2, 4, 6, 8$.

19. It looks pretty clear from these examples that $a^m \equiv 1 \pmod{n}$ only when $(a, n) = 1$. The m situation is not quite so clear though. However, $\phi(n)$ always occurs as one of the m.

Conjecture. *If $(a, n) = 1$, then $a^{\phi(n)} \equiv 1 \pmod{n}$.*

20. This is all just a matter of doing the arithmetic.

21. (i) $x = 3$;   (ii) $x = 4$;   (iii) $x = 3$;   (iv) $x = 4$.

22. (a) If $tx \equiv x \pmod{n}$, then $tx = x + kn$ for some n. Hence $(t-1)x = kn$. Since $(x, n) = 1$, $t-1$ must be a multiple of n. Hence $t \equiv 1 \pmod{n}$.

(b) If $(a, n) = 1$, then by Chapter 1 of *First Step*, page 6, there exist α and β such that $\alpha a + \beta n = 1$. Hence $\alpha a \equiv 1 \pmod{n}$.

If $ay \equiv 0 \pmod{n}$, then $\alpha a y \equiv 0 \pmod{n}$. But $\alpha a \equiv 1 \pmod{n}$, so $y \equiv 0 \pmod{n}$ as required.

(c) Suppose $(ab, n) = t$. Let p be the smallest prime divisor of t. Since $p \mid ab$ (and $p \mid n$), then $p \mid a$ or $p \mid b$ (or both). But this implies that $(a, n) \geq p$ or $(b, n) \geq p$. This contradiction implies that $(ab, n) = 1$.

23. (a) $T = \{1, 3, 5, 7\}$. Now check all possibilities. (For instance $3 \times 1 \equiv 3$, $3 \times 3 \equiv 1, 3 \times 5 \equiv 7, 3 \times 7 \equiv 5 \pmod 8$.)

   The x' set equals T, again by looking at all possibilities.

   (b) Now $(1 \times t)(3 \times t)(5 \times t)(7 \times t) = 1 \times 3 \times 5 \times 7 \pmod 8$ from (a) because tx is congruent to something in T and $tx_1 \not\equiv tx_2 \pmod 8$ unless $x_1 = x_2$. Hence $t^4(1 \times 3 \times 5 \times 7) \equiv 1 \times 3 \times 5 \times 7 \pmod 8$. But $1 \times 3 \times 5 \times 7 \equiv 1 \pmod 8$. Hence $t^4 \equiv 1 \pmod 8$ for all $t \in T$.

24. (a) This can be done by trial and error. Can you find a better way?

   (b) Use the technique of Exercise 23(b).

25. Have a go. Use the last two exercises as a model. Then look at the text for confirmation.

26. Not always. Have a look at Exercise 1.

27. If $n = p$, a prime, then $\phi(p) = p - 1$. Now if $1 \le a < p$, then $(a, p) = 1$. Hence by Euler's Theorem $a^{\phi(p)} = a^{p-1} \equiv 1 \pmod p$.

28. (i) $3^6 \equiv 1 \pmod 7$ ∴ $3^{56} = 3^{54}3^2 \equiv 3^2 \pmod 7$. Hence $b = 2$;

   (ii) $3^{10} \equiv 1 \pmod{11}$. Hence $b \equiv 3^6 \pmod{11}$. So $b = 3$;

   (iii) $b = 0$;

   (iv) $b = 7^2 \equiv 5 \pmod{11}$;

   (v) $b = 5$;

   (vi) $b = 10$.

29. Now if $p = 2, 2p - 1 = 3$ and $\phi(2p) = 2$. Hence $a^{2p-1} = a^3 = a^2 a \equiv a \pmod{2p}$, since $a^2 \equiv 1 \pmod{2p}$. (Check for $a = 1$ and 3.)

   If $p > 2$, then $\phi(2p) = p - 1$. Hence $a^{2p-1} = a^p a^{p-1} \equiv a^p \pmod{2p}$. But $a^p = a^{p-1}a$. So $a^p \equiv a \pmod{2p}$. This means $a^{2p-1} \equiv a \pmod{2p}$.

   If $a^{n-1} \equiv a \pmod n$, then $a^{n-2} \equiv 1 \pmod n$ using Exercise 22(a) with $(a, n) = 1$. Now $a^{n-2} \equiv 1 \pmod n$ for all n for which $\phi(n)$ divides $n - 2$. We know from above that this is true for $n = 2p$ when p is a prime. Are there other possibilities for n?

   Suppose $n = p_1^{\alpha_1} p_2^{\alpha_2} \cdots p_r^{\alpha_r}$ where the p_i are odd primes and $p_1 < p_2 < \cdots < p_r$ and $r \ge 1$. Then we need $p_1^{\alpha_1} p_2^{\alpha_2} \cdots p_r^{\alpha_r} - 2 = k[p_1^{\alpha_1-1} p_2^{\alpha_2-1} \cdots p_r^{\alpha_r-1}(p_1 - 1)(p_2 - 2) \cdots (p_r - 1)]$ for some $k \in N$. Since p_1 is odd, $p_1 - 1$ is even, so the right-hand side is even. However, the left side is odd, so we have a contradiction. This means that n is divisible by 2.

   Suppose $n = 2^\alpha$ for some $\alpha \in N$. Then $n - 2 = 2^\alpha - 2 = 2(2^{\alpha-1} - 1)$. Further $\phi(n) = 2^{\alpha-1}$. We thus require $2(2^{\alpha-1} - 1) = k2^{\alpha-1}$ for some $k \in N$. Hence $2^{\alpha-1} - 1 = 2^{\alpha-2}k$. For $\alpha > 2$ the left-hand side of this last equation is odd and the right-hand side is even. That means $\alpha = 1$ or 2.

Since $\phi(2) = 1$ and $a^{2-1} = a$ we have $a^{n-1} \equiv a \pmod 2$ for $n = 2$. We have already dealt with the case $n = 4$ earlier. So we know that $a^{4-2} \equiv 1 \pmod 4$.

Now suppose $n = 2^\alpha m$, where m is odd and greater than one. Since $\phi(n) = 2^{\alpha-1}\phi(m)$ we require $2^\alpha m - 2 = k(2^{\alpha-1}\phi(m))$. If $\alpha > 1$, then $2^{\alpha-1}m - 1 = k(2^{\alpha-2}\phi(m))$. But this is a contradiction since the left side is odd and the right side is even, since $\phi(m)$ is even.

If $\alpha = 1, 2m - 2 = k\phi(m)$, so $m - 1 = k(1/2\phi(m))$. Let $m = p_1^{\alpha_1} p_2^{\alpha_2} \cdots p_r^{\alpha_r}$, where the $\alpha_i \in N$, the p_i are prime and $p_1 < p_2 < \cdots < p_r$, and $r \geq 1$.

Then

$$p_1^{\alpha_1} p_2^{\alpha_2} \cdots p_r^{\alpha_r} - 1 = \frac{k}{2}[p_1^{\alpha_1-1} p_2^{\alpha_2-1} \cdots p_r^{\alpha_r-1}(p_1 - 1)(p_2 - 1) \cdots (p_r - 1)].$$

If $\alpha_i > 1$ for some i, then p_i divides the right-hand side but not the left-hand side. Hence $\alpha_1 = \alpha_2 = \cdots = \alpha_r = 1$.

We thus have $p_1 p_2 \cdots p_r - 1 = \frac{k}{2}(p_1 - 1)(p_2 - 1) \cdots (p_r - 1)$.

It's at this point that I threw up my hands in horror because I could make no progress. Certainly if $r = 1$ the equation holds. But does it hold for $r > 1$?

Let $M = p_1 p_2 \cdots p_r$, where the p_i are distinct odd primes. Then we want to find M such that $M - 1 = k'\phi(M)$. More generally we want to know if $\phi(M)$ divides $M - 1$. In Richard Guy's book "*Unsolved Problems in Number Theory*" (published by Springer), he states that "D.H. Lehmer (a famous contemporary number theorist) has conjectured that there is no composite value of n such that $\phi(n)$ is a divisor of $n - 1$". So nobody knows of any $r \geq 2$ for which $p_1 p_2 \cdots p_r - 1 = \frac{k}{2}(p_1 - 1)(p_2 - 1) \cdots (p_r - 1)$. That makes me feel a little better.

30. (a) Since $2730 = 2 \times 3 \times 5 \times 7 \times 13$, $\phi(2730) = 1 \times 2 \times 4 \times 6 \times 12 = 576$. Certainly then $a^{576} - 1 \equiv 0 \pmod{2730}$. So $a^{577} - a \equiv 0 \pmod{2730}$.

That doesn't seem to be getting us anywhere so let's have a look at the prime factors of 2730.

Now $\phi(13) = 12$ so $a^{12} \equiv 1 \pmod{13}$ which means that $a^{13} \equiv a \pmod{13}$. This is strictly speaking true for all a. Euler's Theorem gives $a^{12} \equiv 1 \pmod{13}$ for all a relatively prime to 13. Clearly if a is not relatively prime to 13, then $a^{13} \equiv a \pmod{13}$ since $a \equiv 0 \pmod{13}$.

Similarly, since $\phi(7) = 6, a^6 \equiv 1 \pmod{7}$ and so $a^7 \equiv a \pmod{7}$ which gives $a^{13} \equiv a \pmod{7}$. By the discussion with 13, this last congruence is true for all a.

So we have $a^{13} = a + 7s$ and $a^{13} = a + 13t$. Hence $7s = 13t$. So 13 must divide s (and 7 must divide t). Consequently $7s$ and $13t$ are multiples of 91. Hence $a^{13} \equiv a \pmod{91}$.

Can we extend this argument? Since $\phi(5) = 4$ we know that $a^4 \equiv 1 \pmod{5}$ so $a^5 \equiv a \pmod{5}$. But then $a^{13} = a^5 \cdot a^4 \cdot a^4 \equiv a^5 \equiv a \pmod{5}$.

So $a^{13} = a + 5r$. We know from above that $a^{13} = a + 91n$. Since $5r = 91n, r$ must be a multiple of 91. Hence $a^{13} \equiv a \pmod{455}$.

Continuing in this way we see that $a^{13} \equiv a \pmod 2$ and $a^{13} \equiv a \pmod 3$. Hence $a^{13} \equiv a \pmod{2730}$.

(b) The divisors of 8160 are $2^5, 3, 5, 17$. Using (a) and the fact that $\phi(2^5) = 16, \phi(3) = 2, \phi(5) = 4$ and $\phi(17) = 16$ we can complete the argument.

31.   (i) $x_1 = 1, x_2 = 2 \therefore r_3 = 2$;

   (ii) $x_1 = 1, x_2 = 3 \therefore r_4 = 3$;

  (iii) $r_5 = 4$;

  (iv) $r_6 = 5$;

   (v) $r_7 = 6$;

  (vi) $r_8 = 1$ (you'd better check this, it doesn't seem to fit the pattern);

 (vii) $r_9 = 8$ (ah, that's better);

(viii) $r_{10} = 9$;

  (ix) $r_{11} = 10$.

32. It looks as if $r_p = p - 1 \equiv -1 \pmod p$. In fact at the moment, $n = 8$ seems to be the only number of any type for which $r_n \not\equiv -1 \pmod n$.

How did your proof go?

33. Maybe $r_n \equiv -1 \pmod n$ except for $n = 8$. Are there other values of n for which $r_n \equiv 1 \pmod n$? Check out 15 and 21.

It turns out that $r_n \equiv 1 \pmod n$ more often than not. $r_n \equiv -1 \pmod n$ for $n = 4, p^\alpha$ or $2p^\alpha$, where p is an odd prime. For all other $n, r_n \equiv 1 \pmod n$.

Here we have a genuine generalisation of Wilson's Theorem. A proof of this result can be found in G.H. Hardy and E.M. Wright "*An Introduction to the Theory of Numbers*", Oxford, 1979.

34. $66 \times 66 \equiv 1 \pmod{67}$ and $1 \times 1 = 1 \pmod{67}$.

If $a^2 \equiv 1 \pmod{67}$, then either $a \equiv 1 \pmod{67}$ or $a \equiv -1 \pmod{67}$ because 67 is a prime. In the former case $a = 1$ and in the latter $a = 66$.

Since $(r, 67) = 1$, then there exists a_r such that $ra_r + 67t = 1$. Hence $ra_r \equiv 1 \pmod{67}$. So for any r there exists a_r such that $ra_r \equiv 1 \pmod{67}$. However $r \neq a_r$ unless $r = 1$ or 66.

35. Use the argument of the last exercise. Then $a \equiv 1 \pmod{53}$ gives $a = 1$ and $a \equiv -1 \pmod{53}$ gives $a = 52$.

Since $(r, 53) = 1$, then there exists a_r such that $ra_r + 53t = 1$. Hence $ra_r \equiv 1 \pmod{53}$. But $r \neq a_r$ unless $r = 1$ or 52.

36. Suppose $r^2 \equiv 1 \pmod{p}$. Then either $r = 1$ or $r = p - 1$ (by Exercise 34). By Theorem 1 of Chapter 1 of *First Step*, $ra_r + pt = 1$. Hence $ra_r \equiv 1 \pmod{p}$. Hence for all r such that $(r, p) = 1$, there exists a_r such that $ra_r \equiv 1 \pmod{p}$. The two numbers r and a_r are distinct unless $r = 1$ or $p - 1$.

37. $(2 - 1)! = 1 \equiv 1 \pmod{2}$.

38. (a) Since $p \equiv 1 \pmod{4}$, then $p = 4s + 1$ for some s. By Wilson's Theorem $(4s)! \equiv -1 \pmod{p}$.

Now
$$
\begin{aligned}
(4s)! &= (2s)!(2s + 1)(2s + 2) \cdots (2s + 2s) \\
&= (2s)!(p - 2s)[p - (2s - 1)] \cdots (p - 1) \\
&\equiv (2s)!(-2s)[-(2s - 1)] \cdots (-1) &&\pmod{p} \\
&\equiv (2s)^{2s} &&\pmod{p} \\
&\equiv (2s)!(2s)! &&\pmod{p} \\
&\equiv [(2s)!]^2 &&\pmod{p} \\
&\equiv [(\tfrac{p-1}{2})!]^2 &&\pmod{p}
\end{aligned}
$$

(b) In this case $p = 4s + 3$ and $(4s + 2)! \equiv -1 \pmod{p}$. Now $(4s + 2)! = (2s + 1)!(2s + 2) \cdots (2s + 2s + 2) = (2s + 1)![p - (2s + 1)] \cdots (p - 1) \equiv [(2s + 1)!]^2(-1)^{2s+1} \pmod{p}$.

But $(-1)^{2s+1} = -1$, so $1 \equiv [(2s + 1)!]^2 \equiv [(\tfrac{p-1}{2})!]^2 \pmod{p}$.

39. Wilson's Theorem tells us that if p is an odd prime then $(p - 1)! \equiv -1 \pmod{p}$. So assume that $(n - 1)! \equiv -1 \pmod{n}$. Clearly n isn't 2 by Exercise 37. I aim now to show that n is an odd prime. So assume that n is not prime.

Let b be a factor of n and $1 < b < n$. Hence $b \mid (n - 1)!$.

Since $(n-1)! \equiv -1 \pmod{n}$ then, for some integer k, $(n-1)!+1 = kn$. But $b \mid n$ and $b \mid (n - 1)!$, so b must also divide 1. This contradicts the fact that b is greater than 1. Hence n must be an odd prime.

41. There is actually a stronger result than the one you've been trying to prove. This is that

$$(a^4)^{15x+1} \equiv a^4 \pmod{N} \text{ for all } a, x \geq 1 \text{ if and only if } N \mid 6814407600.$$

First note that $6814407600 = 2^4 \cdot 3^2 \cdot 5^2 \cdot 7 \cdot 11 \cdot 13 \cdot 31 \cdot 61$ and that

$\phi(3^2) = 6, \phi(5^2) = 20, \phi(7) = 6, \phi(11) = 10, \phi(13) = 12, \phi(31) = 30, \phi(61) = 60$. Let $N = 6814407600$. To show the given congruence is true for all $a, x \geq 1$, we can show

$$a^{60x} a^4 \equiv a^4 \pmod{n} \tag{1}$$

for $n = 2^4, 3^2, 5^2, 7, 11, 13, 31$ and 61 (the prime power divisors of N).

Begin with $n = 2^4$. For a even, a^4 is divisible by 16, so both sides of (1) are congruent to 0. For a odd, $a^4 \equiv 1 \pmod{16}$, so $a^{60x} \equiv 1$, so both sides of (1) are congruent to 1.

Now let n be any of the other possibilities. If a is relatively prime to n, then $a^{\phi(n)} \equiv 1 \pmod{n}$, and recall that $\phi(n) \mid 60$. Thus $a^{60x} \equiv 1$, so each side of (1) becomes a^4. If $(a, n) > 1$ and n is prime, then $n \mid a$ so both sides of (1) become 0. If $(a, n) > 1$ and $n = 3^2$ (or 5^2), then 3 (or 5) divides a, so 3^4 (or 5^4) divides a^4, and both sides of (1) are congruent to 0.

Thus $a^{60x} \cdot a^4 \equiv a^4 \pmod{6814407600}$, and hence modulo any divisor of 681407600.

Now assume that

$$a^{60x} \cdot a^4 \equiv a^4 \pmod{N} \tag{2}$$

for all $a, x \geq 1$.

If $N = 2^i m$ with m odd, then $i \leq 4$; for if $i \geq 5$ then $32 \mid N$, so putting $a = 2, x = 1$ in (2) yields $2^{64} \equiv 2^4 \pmod{32}$, which is not true. So N can have no more 2's than does 6814407600.

Let d be an odd divisor of N. Then by (2) with $a = 2, x = 1$ we have $2^{64} \equiv 2^4 \pmod{N}$, so $2^{60} \equiv 1 \pmod{d}$. Hence

$$d \mid 2^{60} - 1 = 3^5 \cdot 5^2 \cdot 7 \cdot 11 \cdot 13 \cdot 31 \cdot 41 \cdot 61 \cdot 151 \cdot 331 \cdot 1321.$$

In particular there is the possibility that d could be 41, say. But if it were, we would have (from (2) with $a = 3, x = 1$) $3^{64} \equiv 3^4 \pmod{N}$, so $3^{60} \equiv 1 \pmod{41}$. But actually $3^{60} = (3^4)^{15} \equiv (-1)^{15} = -1 \pmod{41}$. Similarly, d cannot be 151, 331, or 1321. Therefore d must divide $3^2 \cdot 5^2 \cdot 7 \cdot 11 \cdot 13 \cdot 31 \cdot 61$, so N must be a divisor of 6814407600.

42. (a) Now $2^1 \equiv 2 \pmod{7}$, $2^2 \equiv 4 \pmod{7}$ and $2^3 \equiv 1 \pmod{7}$. In general, if $n = 3k$, then $2^n = (2^3)^k \equiv 1^k \pmod{7}$, so $2^{3k} - 1 \equiv 0 \pmod{7}$. If $n = 3k + 1$, then $2^n = 2^{3k}2 \equiv 2 \pmod{7}$ while if $n = 3k + 2$, then $2^n \equiv 4 \pmod{7}$. Hence $2^n - 1$ is divisible by 7 if and only if $n = 3k$.

   (b) Now $2^{3k} + 1 \equiv 1 + 1 \pmod{7}$, $2^{3k+1} + 1 \equiv 3 \pmod{7}$ and $2^{3k+2} + 1 \equiv 5 \pmod{7}$. Hence $2^n + 1$ is never divisible by 7.

43. Let $t = qm + r$, where $0 \leq r < m$. Then $a^t = a^{qm} a^r \equiv a^r$ (mod n). Hence $a^r \equiv 1$ (mod n).

    Since m is the smallest number for which $a^m \equiv 1$ (mod n), then $r = 0$. Hence $t = qm$.

44. Since 1978^n and 1978^m agree in their last three digits, the difference $1978^n - 1978^m = 1978^m(1978^{n-m} - 1)$ is divisible by $10^3 = 2^3 \cdot 5^3$. Since the second factor above is odd, 2^3 divides the first. Also $1978^m = 2^m \cdot 989^m$ so $m \geq 3$.

    We can write $m + n = (n - m) + 2m$. To minimise this sum, we take $m = 3$ and seek the smallest value of $d = n - m$, such that $1978^d - 1$ is divisible by $5^3 = 125$, i.e. $1978^d \equiv 1$ (mod 125).

    By Euler's Theorem, $1978^{\phi(125)} \equiv 1$ (mod 125). Hence, by Exercise 43, $d \mid \phi(125)$. Since $\phi(125) = 100, d$ must be a factor of 100.

    If $1978^d - 1$ is divisible by 125, then $1978^d - 1$ is divisible by 5. Hence $1978^d \equiv 3^d \equiv 1$ (mod 5). The smallest value for d in the congruence $3^d \equiv 1$ (mod 5) is 4. Hence d must be a multiple of 4.

    So $d = 4$, 20 or 100.

    Now $1978^4 \equiv 6$ (mod 125) and $1978^{20} \equiv 26$ (mod 125). Hence $d = n - m = 100$ and the smallest value of $n + m$ is 106.

45. Now $(a + b)^7 - a^7 - b^7 = 7ab(a + b)(a^2 + ab + b^2)^2$.

    Since 7 does not divide $ab(a + b)$, we must choose a, b so that 7^3 divides $a^2 + ab + b^2$, i.e. so that $a^2 + ab + b^2 \equiv 0$ (mod 7^3).   (1)

    Since $a^3 - b^3 = (a - b)(a^2 + ab + b^2)$

    (1) is equivalent to $a^3 \equiv b^3$ (mod 7^3).

    Now $\phi(7^3) = 3 \times 98$. Hence $(a^3)^{98} \equiv 1$ (mod 7^3) for $(a, 7) = 1$. Let $b = 1$ and $a = 2^{98}$, then $(2^{98})^3 \equiv 1$ (mod 7). Since $2^{98} \equiv 4$ (mod 7), $a + b = 2^{98} + 1 \equiv 5$ (mod 7) and $a - b = 2^{98} - 1 \equiv 3$ (mod 7), 7 does not divide $ab(a + b)$ nor $a - b$. So $a = 2^{98}$ and $b = 1$ is a pair of the required form.

    (Actually $a = 18$ and $b = 1$ is a smaller pair. Check that this pair is indeed a solution to the problem.)

46. Consider the set of numbers $1, 2, \ldots, 2p$, where $p > 3$ is a prime. We can choose i numbers from the set $\{1, 2, \ldots, p\}$ and $p - i$ numbers from the set $\{p + 1, p + 2, \ldots, 2p\}$. This will give us p numbers in all. There are $\binom{p}{i} \binom{p}{p - i}$ ways of doing this where i takes the value $0, 1, 2, \ldots, p$. Therefore

$$\binom{2p}{p} = \sum_{i=0}^{p} \binom{p}{i} \binom{p}{p - i} = \sum_{i=0}^{p} \binom{p}{i}^2 = \sum_{i=1}^{p-1} \binom{p}{i}^2 + 2.$$

Hence $\binom{2p}{p} - 2 = \sum_{i=1}^{p-1}\binom{p}{i}^2 = \sum_{i=1}^{p-1}[\frac{p!}{i!(p-i)!}]^2 = p^2\sum_{i=1}^{p-1}[\frac{(p-1)!}{i!(p-1)!}]^2$.

Therefore to prove that $p^3 | \binom{2p}{p} - 2$, it is sufficient to prove that

$$p \left| \sum_{i=1}^{p-1} \left[\frac{(p-1)!}{i!(p-i)!}\right]^2 \right. . \tag{1}$$

Result (1) can be proved as follows:

First we note that since $\binom{p}{i} = \frac{p!}{i!(p-1)!}$ is integral, so is $\frac{(p-1)!}{i!(p-1)!}$. This is because $(p, i!(p-i!)) = 1$, as p is prime. Now, by Wilson's Theorem, $(p-1)! \equiv -1 \pmod{p}$ and so $[(p-1)!]^2 \equiv 1 \pmod{p}$. Therefore we may multiply (1) by $[(p-1)!]^2$ to get

$$\sum_{i=1}^{p-1}\left[\frac{(p-1)!}{i!(p-i)!}\right]^2 \equiv [(p-1)!]^2 \sum_{i=1}^{p-1}\left[\frac{(p-1)!}{i!(p-i)!}\right]^2$$

$$\equiv \sum_{i=1}^{p-1}\left[\frac{(p-1)!(p-1)!}{i!(p-i)!}\right]^2 \pmod{p}. \tag{2}$$

Let $t_k = \frac{(p-1)!(p-1)!}{k!(p-k)!}$. Clearly $t_1 = (p-1)! \equiv -1 \pmod{p}$ (Wilson).

Also $t_{k+1} \equiv t_k[\frac{-k}{k+1}] \pmod{p}$. Suppose for some $k, t_k^2 \equiv \frac{1}{k^2} \pmod{p}$. Then $t_{k+1}^2 \equiv t_k^2(\frac{p-k}{k+1})^2 \equiv \frac{1}{k^2}\frac{(-k)^2}{(k+1)^2} \equiv \frac{1}{(k+1)^2} \pmod{p}$.

Therefore since $t_1^2 \equiv (-1)^2 \equiv 1 \pmod{p}$, by induction, $t_k^2 \equiv \frac{1}{k^2} \pmod{p}$.

Hence $\sum_{i=1}^{p-1}\left[\frac{(p-1)!(p-1)!}{i!(p-i)!}\right]^2 \equiv \sum_{i=1}^{p-1}t_i^2 \equiv \sum_{i=1}^{p-1}\frac{1}{i^2} \pmod{p}.$
$$\tag{3}$$

From (1), (2) and (3), $\binom{2p}{p} - 2 \equiv 0 \pmod{p^3}$ ($p \geq 5$ and prime) if and only if

$$\sum_{i=1}^{p-1}\frac{1}{i^2} \equiv 0 \pmod{p}.$$

Now $\frac{1}{i^2} \pmod{p}$ means the inverse of $i \pmod{p}$ squared. Since if p is prime the inverses of the numbers are all distinct, every number (except 0) will appear mod p, as $\frac{1}{i}$ takes various values for $i = 1, 2, \ldots, p-1$. Similarly every number will appear (except 0) as i takes values from 1 to $p-1$. Therefore

$$\sum_{i=1}^{p-1}\frac{1}{i^2} \equiv \sum_{i=1}^{p-1}i^2 \equiv \frac{(p-1)p(2p-1)}{6}\pmod{p}.$$

Since p is a prime number greater than 3, 2 and 3 do not divide p, whence 6 does not divide p. Therefore $\frac{(p-1)p(2p-1)}{6} \equiv 0 \pmod{p}$.

Note that the result does not hold for $p = 2$ or 3.

5. (Continued)   **Conjecture 3.** $a^{n-1} \equiv 0 \pmod{n}$ *if and only if* $n = p_1^{\alpha_1} p_2^{\alpha_2} \cdots p_r^{\alpha_r}$, *where the* p_i *are distinct primes and the* α_i *are natural numbers,* **at least one of which is greater than** 1, *and* $a = p_1^{\beta_1} p_2^{\beta_2} \cdots p_r^{\beta_r}$, *where* $1 \le \beta_i \le \alpha_i$.

Proof. If n is as stated and $a = p_1 p_2 \cdots p_r$ then $a^{n-1} = (p_1 p_2 \cdots p_r)^{n-1} = (p_1^{\alpha_1} p_2^{\alpha_2} \cdots p_r^{\alpha_r})(p_1^{n-1-\alpha_1} p_2^{n-2-\alpha_2} \cdots p_r^{n-1-\alpha_r}) \equiv 0 \pmod{n}$.

If $a^{n-1} \equiv 0 \pmod{n}$, then $n \mid a^{n-1}$. Hence if p is any prime divisor of n, then $p \mid a^{n-1}$ and so $p \mid a$. The result then follows.   □

Incidentally, now you know the above result you should be able to find the smallest number m for which $a^m \equiv 0 \pmod{n}$.

Chapter 5

Means and Inequalities

5.1. Introduction

The theme of this chapter is inequalities, though I take some time at the start to introduce a few examples of averages or means, some of which you may not have seen before. But even in the two sections on means, more effort is spent on the use of inequalities than not.

Here much of the time I have used "means" to beget inequalities to beget more inequalities. In the process I look at possible extensions and generalisations with a view to seeing how far I can take the concepts in a pure mathematical vein. You should notice the mathematical process considered in Chapter 3 coming into play here on a regular basis.

Occasionally in this chapter I will notice an application too. For instance, inequalities are useful to find extreme values of functions, with or without restrictions on the variables, as well as to root out some interesting geometrical facts.

In later life (at university?) you will find inequalities invaluable in calculus and in analysis, the theoretical foundation of calculus. However, I hope that some of the nice inequalities that are presented here make the effort of mastering the underlying material worthwhile. So, what are you waiting for? Read on.

5.2. Rules to Order the Reals By

I've blithely used the symbols $>$, $\geq$, $<$ and $\leq$ throughout both *First Step* and this book, without taking too much notice of the rules that they obey. Since those four little symbols are the basis of the material of this chapter, I'm going to take them a little more seriously than I have done so far. So I'll take some time to investigate $\leq$, the famous less than or equal to sign.

Now $\leq$ provides order to $\mathbb{R}$, the real numbers. Given two real numbers a and b, we know that $a \leq b$ if a is equal to b or lies to the left of b on the real number line or if $b - a \geq 0$.

Exercises

1. Which of the following are true and which are false?
   (i) $53 \leq 64$; (ii) $-37 \leq 53$; (iii) $-37 \leq -64$; (iv) $53 \leq 53$.
2. Determine the following sets.
   (i) $\{x : x \leq 64\} \cap \{y : 64 \leq y\}$;
   (ii) $\{x : 10 \leq x\} \cap \{y : y \leq 12\}$;
   (iii) $\{x : 0 \leq (x - 3)^2\}$;
   (iv) $\{x : (x - 3)^2 \leq 0\}$.
3. (a) If $x \leq y$, is it true that for any real number $r, x + r \leq y + r$?
   (b) If $x \leq y$, is it true that for any real number $r, rx \leq ry$?
   (c) If $x \leq y$, is it true that $x^2 \leq y^2$?

There are eight basic properties that the ordering $\leq$ has when acting on the real numbers. These are listed below. Some of them have names which I've put in brackets. In all cases x, y, z and r are any real numbers subject only to the provisos of the properties.

P1. $x \leq x$. (reflexive)
P2. If $x \leq y$ and $y \leq x$, then $x = y$. (antisymmetric)
P3. If $x \leq y$ and $y \leq z$, then $x \leq z$. (transitive)
P4. One of the following three holds: $x < y$ or $y < x$ or $x = y$. (trichotomy)
P5. If $x \leq y$, then $x + r \leq y + r$.
P6. If $x \leq y$ and $0 \leq r$, then $rx \leq ry$.
P7. If $x \leq y$ and $r \leq 0$, then $ry \leq rx$.
P8. $0 \leq x^2$.

These properties should confirm all you always knew about inequalities and real numbers.

Exercises

4. Which of the eight properties of $\leq$ also hold for $<$?
5. Define the symbol $\square$ on $\mathbb{Z}$, the integers, by $x\square y$ if $x \equiv y$ (mod 63). Which of the properties P1 to P8 inclusive, hold for $\square$?
6. Let S be any collection of sets and let $\subseteq$ be the usual set inclusion. Further, in the properties P1 to P8 above, let $+$ be set union; multiplication be set

intersection; strict containment be strictly less than, and 0 be the empty set; Which of the properties P1 to P8 inclusive, hold for $\subseteq$ and S?

7. Use whichever of properties P1 to P8 that you need to prove the following:
   (i) If $a \le b$ and $c \le d$, then $a + c \le b + d$;
   (ii) If $0 \le a \le b$ and $0 \le c \le d$, then $0 \le ac \le bd$;
   (iii) If $0 \le a \le b$, then $a^2 \le b^2$;
   (iv) If $0 < a \le b$, then $0 < \frac{1}{b} \le \frac{1}{a}$.
8. (i) If $a \le b$ and $c \le d$ is it true that $ac \le bd$?
   (ii) If $a \le b$ is it true that $\frac{1}{b} \le \frac{1}{a}$.
9. Show that P7 and P8 follow from P5 and P6.

5.3. Means Arithmetic and Geometric

As you probably know, there are all sorts of averages. You've got your means, your medians and your modes, and more than likely a lot of other things too. I don't want to worry about the relevance of these to particular situations. (Anyone who wants to know more about this should read Darrell Huff's classic book "How to Lie with Statistics".) What I do want to do though, is to remind you of the arithmetic mean and talk about some mean old means that you may not have met.

So first the arithmetic mean. This is used to find the average height of people in a given class or the average weight of chocolate bars or the average family income. You simply add up all the quantities involved and divide by the number of participants.

The **arithmetic mean** of $a_1, a_2, \ldots, a_n$ is

$$A(n) = \frac{1}{n} \sum_{i=1}^{n} a_i.$$

Exercises

10. In Form 2T, the students have the following weights: 48 kg, 52 kg, 53 kg, 54 kg, 54 kg, 56 kg, 60 kg, 61 kg, 63 kg, 279 kg.
    Find the average weight of the students in Form 2T.
11. Freda cycled 15 km at an average speed of 10 kmh^{-1} and then cycled a further 10 km at 5 kmh^{-1}. What was her overall average speed?
12. A **weighted arithmetic mean** is given by $\dfrac{\sum_{i=1}^{n} w_i x_i}{\sum_{i=1}^{n} w_i}$. Show how this is used to find students averages for a year, when different school projects are given different total scores.
    Show how the weighted mean is used in some sports statistics.

13. The Crockett Cricket Club awards its bowling trophy each year to the bowler who has the best average (computed by dividing the runs hit off the bowler by the wickets taken) over the whole season.

    In 2011–12, Chester Chucker had an average of 50 runs per wicket for the first half of the season and an average of 25 runs per wicket for the second half of the season. On the other hand, Slim (Slow) Spin had a worse average in each half of the season than Chester. Slim's two averages were 51 in the first half and 25.16 in the second.

    (a) Give an example to show that Slim could end up with the Crockett Cricket Club bowling trophy for 2011–12?

    (b) Can Chester ever win the trophy?

14. If $\dfrac{a_1}{b_1} < \dfrac{a_2}{b_2}$ and $\dfrac{c_1}{d_1} < \dfrac{c_2}{d_2}$, is $\dfrac{a_1 + c_1}{b_1 + d_1} < \dfrac{a_2 + c_2}{b_2 + d_2}$, where all the terms are positive?

15. Show that the arithmetic mean of a and b is the term c where a, c, b are in arithmetic progression. (That is, show that $c - a = b - c$.)

16. The **geometric mean** of the non-negative numbers a_1 and a_2 is $\sqrt{a_1 a_2}$. Show that a_1, $\sqrt{a_1 a_2}$ and a_2 are three consecutive terms of a geometric progression. Also show, that if a_1, a_2 are distinct and non-zero, then $\sqrt{a_1 a_2}$ lies strictly between them.

Now I'll generalise the geometric mean of Exercise 16. For non-negative numbers $a_1, a_2, \ldots, a_n$, their **geometric mean** is said to be

$$G(n) = (a_1 a_2 \cdots a_n)^{\frac{1}{n}}.$$

Note that allowing any of the a_i to be negative might force us to try to take an illegal n-th root.

Exercises

17. Find the arithmetic and geometric means of the following sets of numbers.
    (i) $\{2, 4, 8, 16, 32\}$; (ii) $\{3, 9, 27, 81\}$; (iii) $\{2, 2, 8, 25, 125\}$.

18. For the sets in the last exercise, is $G(n)$ bigger than $A(n)$?

19. If a_1, a_2 are non-negative, when is $A(2) \geq G(2)$?

20. If a_1, a_2 are non-negative, when is $A(2) = G(2)$?

21. What evidence do we have for the conjecture that $A(n) \geq G(n)$ for all n? Determine whether the conjecture is true or not. If it is, when does $A(n) = G(n)$?

22. Prove that $(a + b)(b + c)(c + a) \geq 8abc$ for any a, b, $c > 0$. When does equality hold here?

(What tools have you got to attack this animal with? Can it be done with $A(2) \geq G(2)$? Maybe we need to use this inequality more than once.)

23. Show that $x^3 + y^3 + z^3 > 3xyz$ if $x, y, z > 0$.

No hints this time.

24. Show that $\dfrac{yz}{x^2} + \dfrac{zx}{y^2} + \dfrac{xy}{z^2} \geq 3$ if $x, y, z, > 0$.

25. Experiment with and try to extend the arithmetic and geometric means and see what other inequalities you can come up with.

26. Let $s = a + b + c + d$, where $a, b, c, d > 0$ and not all of a, b, c, d are equal. Show that

$$(s - a)(s - b)(s - c)(s - d) > 81abcd.$$

27. Given $a_1, a_2, \ldots, a_r > 0$, let $s = \sum_{i=1}^{n} a_i$. Prove that

$$(1 + a_1)(1 + a_2) \cdots (1 + a_r) \leq \sum_{k=0}^{n} \frac{s^k}{k!}.$$

28. Prove that if θ is an acute angle, then $\tan \theta + \cot \theta \geq 2$. What can be said about $\tan \theta + \cot \theta$ for all other values of θ?

29. Prove that for any integer $n > 1$, $n! < \left(\frac{n+1}{2}\right)^n$.

30. Prove that for any positive numbers $a_1, a_2, \ldots, a_n$,

$$\frac{a_1}{a_2} + \frac{a_2}{a_3} + \cdots + \frac{a_n}{a_1} \geq n.$$

In Exercise 20 we discovered and proved the arithmetic mean-geometric mean inequality. The next few exercises show how to use that inequality to find extreme values of functions without using calculus. Try them now. If they cause you trouble, have a look through Example 1. It may give you a hint as to how to proceed. The new trick that you need is to see how to set the problem up so that the $A(n) \geq G(n)$ inequality can be used.

Example 1. For $0 \leq x \leq 4$, what value(s) of x produces the maximum value of $f(x) = x^3(4 - x)$?

Now I want to use the $A(n) \geq G(n)$ inequality to obtain this maximum. This means that $f(x)$ must be at most $A(n)$ because I am looking for the maximum value of f. But I need to get $A(n)$ as a constant. If it is a function of x, I will have made no progress.

To do this, let $a_1 = \frac{x}{3} = a_2 = a_3$ and $a_4 = 4 - x$. Then $\sum_{i=1}^{4} a_i = 4$, so $A(4) = \frac{4}{4} = 1$. On the other hand, $G(4) = \left[\frac{x^3(4-x)}{27}\right]^{\frac{1}{4}}$.

So $\left[\frac{x^3(4-x)}{27}\right] \leq 1^4 = 1$. Hence $x^3(4 - x) \leq 27$.

So the maximum of 27 is achieved when $A(4) = G(4)$. That is, when $a_1 = a_2 = a_3 = a_4$. So $\frac{x}{3} = (4 - x)$. This is when $x = 3$.

Note how we set up $G(4)$ to be a multiple of $f(x)$, while simultaneously we made $A(4)$ a constant. This kind of manipulation will be useful in what follows.

Exercises

31. Given a, $b > 0$, what values of x minimise $\dfrac{a + bx^4}{x^2}$?

32. For $0 \leq x \leq 9$, what value of x gives $x^2(9 - x)$ its biggest value?

33. Find the maximum value of $(5 + 2x)^3(3 - x)^5$ for $-\frac{5}{2} < x < 3$.

If we are careful with our use of the arithmetic mean-geometric mean inequality, we can also manage extreme values of functions where there are restrictions on the variables. This next set of exercises contains problems of that kind. The trick with these max–min problems is to split the function up so that the appropriate part of the inequality is constant. In other words, use a similar method to Example 1.

Exercises

34. Prove that for any $n \in \mathbb{N}$ and for a, $b \geq 0$, $(ab^n)^{\frac{1}{n+1}} \leq \dfrac{a + nb}{n + 1}$.

35. Find the greatest value of x^2y^3z given that $x, y, z \geq 0$ and $x + y + z = c$, where c is constant.

36. Find the minimum value of $x + y + z$, given that $x^2y^3z = 108$ and x, y, $z \geq 0$. For what values of x, y, z is this minimum obtained?

37. For what values of x, y, $z \geq 0$ does x^3y^2z achieve its greatest value, given that $6x + 4y + z = 1$?

We end this section with some geometrical applications.

Exercises

38. Prove that among all rectangles with a given perimeter P, the square has greatest area.

39. Prove that among all rectangles with a given area A, the square has shortest perimeter.

40. Find the volume of a rectangular box in terms of the areas x, y, z of three different faces which meet at a vertex. (Assume the box is closed.)
    Prove that for a given surface area A, the volume V is greatest when the box is a cube.

41. Prove that of all triangles with a given perimeter P, the equilateral triangle has greatest area.

5.4. More Means

The next mean I want to mention is the harmonic mean, $H(n)$. For non-zero $a_1, a_2, \ldots, a_n$, the **harmonic mean** is defined to be

$$H(n) = \frac{n}{\frac{1}{a_1} + \frac{1}{a_2} + \cdots + \frac{1}{a_n}}.$$

Clearly we can't have $a_i = 0$. So we need the "non-zero" restriction.

This definition though, might seem a little perverse. However, you first have to realise that the **harmonic series** is $1 + \frac{1}{2} + \frac{1}{3} + \frac{1}{4} + \cdots$, so it does make sense using the term "harmonic mean" for something with $\frac{1}{a_1} + \frac{1}{a_2} + \cdots + \frac{1}{a_n}$ in it.

But why take $H(n)$ to be the reciprocal of the more obvious "mean": $\frac{\frac{1}{a_1} + \frac{1}{a_2} + \cdots + \frac{1}{a_n}}{n}$? We'll get to that later.

Exercises

42. Show that if $a_i = a$ for $i = 1, 2, \ldots,$ where a is positive, then $A(n) = G(n) = H(n) = a$.
43. Show that if $\frac{1}{a} - \frac{1}{b} = \frac{1}{c} - \frac{1}{b}$, then b is the harmonic mean of a and c.

The observations of Exercises 42 and 43 begin to make sense of the definition of $H(n)$. If I had taken the "obvious" mean suggested before Exercise 42, $H(n)$ would not have been consistent with the other means defined so far. It also would not have been consistent with the arithmetic, geometric and harmonic series properties. Hence we stick with the perverse expression

$$H(n) = \frac{n}{\frac{1}{a_1} + \frac{1}{a_2} + \cdots + \frac{1}{a_n}}.$$

Exercises

44. Find the harmonic mean of the sets of numbers in Exercise 17.
45. Is $G(2) \leq H(2)$ or is $G(2) \geq H(2)$? When is $G(2) = H(2)$?
46. Use the fact that $A(3) \geq G(3)$ to prove that $G(3) \geq H(3)$ when $a_1, a_2, a_3 > 0$. When does equality hold?

47. Prove that $G(n) \geq H(n)$ when the a_i are all positive. When is $G(n) = H(n)$?

48. Prove that $(a + b + c)\left(\frac{1}{a+b} + \frac{1}{b+c} + \frac{1}{c+a}\right) \geq \frac{9}{2}$.

49. If a, b, c are positive real numbers, show that $\frac{a}{a+b} + \frac{b}{c+a} + \frac{c}{a+b} \geq \frac{3}{2}$. When does equality hold?

50. Prove necessary and sufficient conditions for the harmonic mean of two numbers to equal their arithmetic mean.

51. Prove that the geometric mean of a pair of positive numbers is the geometric mean of their harmonic and arithmetic means.

The next mean I want to mention is the **quadratic mean**, $Q(n)$. For $a_1, a_2, \ldots, a_n$, I'll define

$$Q(n) = \sqrt{\frac{a_1^2 + a_2^2 + \cdots + a_n^2}{n}}.$$

Note that $Q(n)$ is defined for all real a_i. There is no need to restrict them to being non-negative.

Exercises

52. Show that $Q(n) = a$ if $a_i = a$ for $i = 1, 2, \ldots, n$. (So, with respect to that property, $Q(n)$ is consistent with all other means so far defined.)

53. Where does $Q(2)$ fit in the hierarchy $A(2) \geq G(2) \geq H(2)$? Does this generalise?

54. Let the sides of a right angled triangle be a, b, c, where c is the hypotenuse. Prove that $a + b \leq c\sqrt{2}$.

55. Prove that, if $a + b = 1$, where a, $b > 0$, then

$$\left(a + \frac{1}{a}\right)^2 + \left(b + \frac{1}{b}\right)^2 \geq \frac{25}{2}.$$

56. Show that $(x + y)^2 + (y + z)^2 + (z + x)^2 \leq 4(x^2 + y^2 + z^2)$.

57. Let $f(x, y, z) = 12x + 3y + 4z$. Find the maximum value of f on the surface of the sphere $x^2 + y^2 + z^2 = 1$.

At what point(s) is this maximum attained? (The $12x$, $3y$ and $4z$ need to be treated in a way that will lead to $x^2 + y^2 + z^2$. See Exercise 35 for a hint.)

Now at the moment all four means that we have talked about look quite unrelated, yet there is a general pattern. Consider what is called the *generalised f-mean*:

$$M_f = f^{-1}\left[\frac{1}{n}f(a_1) + \frac{1}{n}f(a_2) + \cdots + \frac{1}{n}f(a_n)\right]$$

$$= f^{-1}\left(\frac{1}{n}\sum_{i=1}^{n} f(a_i)\right),$$

where f is any old one-to-one function. Here f^{-1} is the function that "undoes" f, that is, $f(f^{-1}(x)) = x = f^{-1}(f(x))$.

Let's take $f(x) = x$ as an example to see what is going on here. In fact, for this f, $f^{-1}(x) = x$ too; check it out. Now if I substitute for f and f^{-1} in the formula for M_f, I get

$$M = f^{-1}\left[\frac{1}{n}a_1 + \frac{1}{n}a_2 + \cdots + \frac{1}{n}a_n\right] = \left[\frac{1}{n}a_1 + \frac{1}{n}a_2 + \cdots + \frac{1}{n}a_n\right] = A(n).$$

So using $f(x) = x$, M_f gives us the arithmetic mean. It should be possible to get other means that I have introduced earlier as special cases of the f-mean.

Exercises

58. Which of the following functions defined on the real numbers are one-to-one functions?
   (i) $f(x) = x^2$; (ii) $f(x) = x^3$; (iii) $f(x) = x^{1/3}$; (iv) $f(x) = \sin x$.
   Which of the above functions can be made into one-to-one functions by limiting their domain?
59. What means are obtained by letting $f(x)$ be (i) x^2; (ii) $\frac{1}{x}$?
60. How is the geometric mean obtained from M_f?
61. Experiment with various functions f to see what interesting properties your means have.
62. The **power mean**, M_p is defined by letting $f(x) = x^p$.
   (i) If $p > q$, what, if any, is the relationship between M_p and M_q? (You might get a conjecture here by trying small values of p and q. The proof is not easy.)
   (ii) Let M_0 be the limit of M_p as $p \to 0$. What mean that we have used earlier in this chapter is equal to M_0? (Experiment to form a conjecture. Again a proof is not easy.)

5.5. More Inequalities

In this section we see that if you can get in first and define an equality that someone finds useful, you may have it named after you. Let's first turn our thoughts to Chebycheff.

Exercise

63. Let $A = \{a_1, a_2, \ldots, a_n\}$, $B = \{b_1, b_2, \ldots, b_n\}$ and $C = \{a_1b_1, a_2b_2, \ldots, a_nb_n\}$. Further let $\overline{A}, \overline{B}, \overline{C}$ be the arithmetic means of the numbers in the sets A, B, C, respectively. It would be nice if there was some relation between $\overline{A} \times \overline{B}$ and $\overline{C}$.
    (a) Show that $\overline{A} \times \overline{B} \neq \overline{C}$ in general.
    (b) Suppose $n = 2$. Let $a_1 \leq a_2$ and $b_1 \leq b_2$. Show that $\overline{A} \times \overline{B} \leq \overline{C}$.
    (c) Repeat (b) with $n = 3$ and $a_1 \leq a_2 \leq a_3$, $b_1 \leq b_2 \leq b_3$.
    (d) Prove that if A and B are non-decreasing sequences, $\overline{A} \times \overline{B} \leq \overline{C}$.
    (e) Prove that if A and B are non-increasing sequences then $\overline{A} \times \overline{B} \leq \overline{C}$.

The result of Exercise 63 is called the **Chebycheff inequality**. We state it again below.

Let $A = \{a_1, a_2, \ldots, a_n\}$, $B = \{b_1, b_2, \ldots, b_n\}$ be non-increasing or non-decreasing sequences and let $C = \{a_1b_1, a_2b_2, \ldots, a_nb_n\}$. Then

$$\overline{A} \times \overline{B} \leq \overline{C}.$$

Exercises

64. Prove that, if $x, y \geq 1$, then

$$\sum_{k=0}^{n} x^k \sum_{k=0}^{n} y^k \leq (n+1) \sum_{k=0}^{n} x^k y^k.$$

What happens if $0 < x, y < 1$?
What happens if $0 < x < 1 \leq y$?
65. When does equality hold in the Chebycheff inequality?

The next inequality that I want to mention is due to Cauchy and Schwarz and Buniakowski. This result was first proved by Cauchy and then generalised, independently, by Schwarz and Buniakowski. Hence, in some circles, the inequality below is referred to as the Cauchy–Schwarz–Buniakowski (or the C-S-B) inequality. For some reason Buniakowski's name is often dropped, so you may just hear of this as the Cauchy–Schwarz inequality.

Exercises

66. Show that $(a_1 b_1 + a_2 b_2)^2 \leq (a_1^2 + a_2^2)(b_1^2 + b_2^2)$.
67. Show that $(a_1 b_1 + a_2 b_2 + a_3 b_3)^2 \leq (a_1^2 + a_2^2 + a_3^2)(b_1^2 + b_2^2 + b_3^2)$.

68. State and prove the generalisation of the last two exercises.

As a result of the last exercise we know that

$$\left(\sum\nolimits_{i=1}^{n} a_i b_i\right)^2 \leq \sum\nolimits_{i=1}^{n} a_i^2 \sum\nolimits_{i=1}^{n} b_i^2.$$

Consequently we can take the square root of both sides to give

$$\left|\sum\nolimits_{i=1}^{n} a_i b_i\right| \leq \left(\sum\nolimits_{i=1}^{n} a_i^2\right)^{\frac{1}{2}} \left(\sum\nolimits_{i=1}^{n} b_i^2\right)^{\frac{1}{2}}.$$

This is the inequality that is usually referred to as the **Cauchy–Schwarz–Buniakowski inequality**. I have added the modulus signs on the left to strengthen the inequality. Clearly without these, if the expression on the left was negative then the inequality would hold anyway.

Exercise

69. When does equality hold in the Cauchy–Schwarz–Buniakowski inequality?

The next example will give you some idea of the uses of the Cauchy–Schwarz–Buniakowski inequality.

Example 2. Let $f(x, y, z) = 3x + y + 2z$. Find the maximum value of f given that $x^2 + y^2 + z^2 = 1$.

Now I want to use the Cauchy–Schwarz–Buniakowski inequality here. Somehow then, I've got to work with the function and its restriction so that I can apply this inequality. Hence I'll try to get $f(x, y, z)$ as the sum of a product of terms from a sequence. At the same time I'll try to get the restriction as the sum of squares and hope I can force another sum of squares to be a constant.

Now the simplest way to get $\sum a_i^2 = x^2 + y^2 + z^2$ is to let $a_1 = x$, $a_2 = y$ and $a_3 = z$.

In this case I need $xb_1 + yb_2 + zb_3 = 3x + y + 2z$. So choose $b_1 = 3$, $b_2 = 1$, $b_3 = 2$. Then $\sum b_i^2 = 14$. By Cauchy–Schwarz–Buniakowski, I get

$$\sum\nolimits_{i=1}^{3} a_i b_i = 3x + y + 2z \leq \left(\sum\nolimits_{i=1}^{3} a_i^2\right)^{\frac{1}{2}} \left(\sum\nolimits_{i=1}^{3} b_i^2\right)^{\frac{1}{2}}$$

$$= (x^2 + y^2 + z^2)^{\frac{1}{2}} \sqrt{14}.$$

Hence $3x + y + 2z \leq \sqrt{14}$.

To find out when this maximum value is achieved, I first recall that I have equality in Cauchy–Schwarz–Buniakowski when $a_i = \lambda b_i$ or $b_i = \mu a_i$ for $i = 1, 2, \ldots, n$ and λ, μ real.

In the present example this means that $x = 3\lambda$, $y = \lambda$, $z = 2\lambda$ (or $3 = \lambda x$, $\lambda = y$, $2 = \lambda z$).

Then $3(3\lambda) + \lambda + 2(2\lambda) = \sqrt{14}$, so $\lambda = \frac{1}{\sqrt{14}}$ and $x = \frac{3}{\sqrt{14}}, y = \frac{1}{\sqrt{14}}$, $z = \frac{2}{\sqrt{14}}$, (or $3\left(\frac{3}{\lambda}\right) + \frac{1}{\lambda} + 2\left(\frac{2}{\lambda}\right)$, which gives the same result).

Exercises

70. Let $f(x, y, z) = x + 2y + 4z$. Find the maximum value of f on the surface given by $x^2 + 2y^2 + z^2 = 1$. Where does the minimum occur?
71. Find the minimum value of $f(x, y, z) = x^2 + y^2 + z^2/2$ given that $x + y + z = 10$. Where does the minimum occur?
72. Solve Exercise 57 using the Cauchy–Schwarz–Buniakowski inequality.

The next inequality I introduce is due to Hölder, it is in fact a generalisation of the Cauchy–Schwarz–Buniakowski inequality. If a_i, b_i, $i = 1, 2, \ldots, n$ are non-negative, then

$$\sum_{i=1}^{n} a_i b_i \le \left(\sum_{i=1}^{n} a_i^p\right)^{\frac{1}{p}} \left(\sum_{i=1}^{n} b_i^q\right)^{\frac{1}{q}},$$

where p, q are real numbers greater than 1 such that $\frac{1}{p} + \frac{1}{q} = 1$.

Exercises

73. For what values of p and q does the Hölder inequality become the C-S-B inequality?
74. Show that if $\lambda a_i^p = \mu b_i^q$ for $i = 1, 2, \ldots, n$ and λ, μ are non-negative, then equality holds in the Hölder inequality.

In Chapter 7 of *First Step*, when we were investigating the modulus function we discovered that

$$|a + b| \le |a| + |b|.$$

This inequality was referred to as the triangle inequality since the sum of two sides of a triangle is greater than the third side.

If we note that $|a| = \sqrt{a^2}$, then the triangle equality can be rewritten as

$$\sqrt{(a + b)^2} \le \sqrt{a^2} + \sqrt{b^2}.$$

This then generalises to the inequality below which we shall also call the triangle inequality.

$$\left(\sum_{i=1}^{n}(a_i+b_i)^2\right)^{\frac{1}{2}} \le \left(\sum_{i=1}^{n}a_i^2\right)^{\frac{1}{2}} + \left(\sum_{i=1}^{n}b_i^2\right)^{\frac{1}{2}}.$$

To see why this has something to do with triangles. Consider the points (a_1, a_2), (b_1, b_2) and (a_1+b_l, a_2+b_2) in the plane. In the diagram, O is the origin.

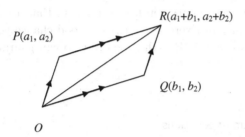

Now $OPRQ$ is a parallelogram, so $QR = OP$. Hence $QR = \sqrt{(a_1^2 + a_2^2)}$, $OQ = \sqrt{(b_1^2 + b_2^2)}$, and $OR = \sqrt{(a_1+b_1)^2 + (a_2+b_2)^2}$, using Pythagoras' Theorem.

But $OQ + QR \ge OR$. Hence we have the triangle inequality for $n = 2$.

Exercises

75. If $A = (a_1, a_2, \ldots, a_n)$ then define $|A|$ to be $\left(\sum_{i=1}^{n}a_i^2\right)^{\frac{1}{2}}$. If $B = (b_1, b_2, \ldots, b_n)$, then how must we define $A + B$ in order to have $|A + B| \le |A| + |B|$?
76. Define $A \cdot B$ so that $|A \cdot B| \le |A| \cdot |B|$.
77. When does equality hold in the triangle inequality?
78. Use the C-S-B inequality to prove the triangle inequality.

Minkowski's inequality does a similar thing for the triangle inequality as Hölder did for C-S-B.

Minkowski's inequality is

$$\left[\sum_{i=1}^{n}|a_i+b_i|^p\right]^{\frac{1}{p}} \le \left[\sum_{i=1}^{n}|a_i|^p\right]^{\frac{1}{p}} + \left[\sum_{i=1}^{n}|b_i|^p\right]^{\frac{1}{p}}$$

where p is a real number greater than or equal to 1.

Exercise

79. For what value of p does the Minkowski inequality reduce to the triangle inequality?

5.6. A Collection of Problems

In this section I present a collection of problems from a variety of sources, which rely on means or inequalities for a solution. They are in no particular order. This means that the last might be the easiest or the first the meanest (and there are some mean ones here). It also means that if you need to use the triangle inequality in Exercise n there's a good chance you won't need to use it in Exercise $n + 1$. But then again, who can guarantee anything these days?

Exercises

80. Solve the system of equations

$$x + y + z = a \tag{1}$$
$$x^2 + y^2 + z^2 = b^2 \tag{2}$$
$$xy = z^2 \tag{3}$$

where a and b are constants. Give the conditions that a and b must satisfy so that x, y, z (the solutions of the system) are distinct positive numbers.

(IMO 1961, Question 1)

81. Let a, b, c be the sides of a triangle, and T its area. Prove:

$$a^2 + b^2 + c^2 \geq 4T\sqrt{3}.$$

In what case does equality hold?

(IMO 1961, Question 2)

82. Let ABC be a triangle and D the point on BC where the incircle of $\triangle ABD$ and the excircle (to side DC) of $\triangle ADC$ have the same radius ρ_1. Define ρ_2, ρ_3 analogously. Prove that $\rho_1 + \rho_2 + \rho_3 \geq \frac{9}{4}r$, where r is the inradius of $\triangle ABC$.

(Crux Mathematicorum, 1990, Solution 1391)

83. Suppose that $a_1, a_2, \ldots, a_n$ are real $(n > 1)$ and

$$A + \sum_{i=1}^{n} a_i^2 < \frac{1}{n-1} \left(\sum_{i=1}^{n} a_i \right)^2.$$

Prove that $A < 2a_i a_j$ for $1 \leq i < j \leq n$.

(Putnam Competition No. 38, Question B5)

84. Suppose a, b, c are the sides of a triangle. Prove that
$$a^2(b+c-a) + b^2(c+a-b) + c^2(a+b-c) \le 3abc.$$
(IMO 1964, Question 2)

85. Prove that for all real numbers $x_1, x_2, y_1, y_2, z_1, z_2$ with $x_1 > 0$, $x_2 > 0$, $x_1y_1 - z_1^2 > 0$, $x_2y_2 - z_2^2 > 0$, the inequality
$$\frac{8}{(x_1+x_2)(y_1+y_2)-(z_1+z_2)^2} \le \frac{1}{x_1y_1 - z_1^2} + \frac{1}{x_2y_2 - z_2^2}$$
is satisfied. Give necessary and sufficient conditions for equality.
(IMO 1969, Question 6)

86. Let f and g be real-valued functions defined for all real values of x and y, and satisfying the equation
$$f(x+y) + f(x-y) = 2f(x)g(y)$$
for all x, y.
Prove that if $f(x)$ is not identically zero, and if $|f(x)| \le 1$ for all x, then $|g(y)| \le 1$ for all y.
(IMO 1972, Question 5)

87. Find all real numbers a for which there exist non-negative real numbers x_1, x_2, x_3, x_4, x_5 satisfying the relations
$$\sum_{k=1}^{5} kx_k = a, \quad \sum_{k=1}^{5} k^3 x_k = a^2, \quad \sum_{k=1}^{5} k^5 x_k = a^3$$
(IMO 1979, Question 5)

88. Let a, b and c be the lengths of the sides of a triangle. Prove that
$$a^2 b(a-b) + b^2 c(b-c) + c^2 a(c-a) \ge 0.$$
Determine when equality occurs.
(IMO 1983, Question 6)

89. Prove that $0 \le yz + zx + xy - 2xyz \le 7/27$, where x, y and z are nonnegative real numbers for which $x + y + z = 1$.
(IMO 1984, Question 1)

90. P is a point inside a given triangle ABC. D, E, F are the feet of the perpendiculars from P to the lines BC, CA, AB, respectively. Find all P for which $BC + CA + AB$ is least.
(IMO 1981, Question 1)

5.7. Solutions

1. (i) true; (ii) true; (iii) false; (iv) true.
2. (i) $\{64\}$; (ii) $\{x : 10 < x < 12\}$; (iii) all real numbers; (iv) $\{3\}$.

3. (a) Yes — adding a positive, negative or zero number to each of x and y does not change their relative positions on the number line.

    (b) no — for instance $-4 \leq 3$ but $(-1)(-4) = 4 \geq (-1)(3) = -3$.

    (c) no — for instance $(-4)^2 \geq (3)^2$.

4. P1 is false — it is not true that x is less than x. Since there are no numbers x, y for which $x < y$ and $y < x$, then we say that P2 is vacuously true. That is, it is true, but no numbers satisfy it (and no numbers disobey it). P3 is true. P4 is true. P5 holds for $<$. P6 remains true, as does P7. P8 is false for $x = 0$.

5. P1 is true — any number is congruent to itself modulo 63. P2 is false $64 \equiv 1 \pmod{63}$ and $1 \equiv 64 \pmod{63}$ but $1 \neq 64$. P3 is true. P4 doesn't make any sense in this context. P5 is true as are P6 and P7. P8 is false.

6. P1, P2, P3 are true. P4 is false because two sets A, B may have empty intersection. In that case it is not true that $A \subset B$, $B \subset A$ or $A = B$. P5, P6 are true. P7 is not true, since if $R \subseteq \varnothing$, then $R = \varnothing$. P8 is true, since $\varnothing \subseteq A \cap A$.

7. (i) $a + c \leq b + c$, by P5. $b + c \leq b + d$, by P5. Hence $a + c \leq b + d$ by P3; (ii) $0 \leq ac \leq bc$, by P6. $0 \leq bc \leq bd$ by P6. $0 \leq ac \leq bd$ by P3; (iii) use (ii) with $c = a$ and $d = b$; (iv) by P6 with $r = \frac{1}{ab}$, $0 < a \leq b$. Hence $0 < \frac{1}{b} \leq \frac{1}{a}$.

8. (i) No : $-2 \leq 1$ and $-3 \leq 2$, but $(-3)(-2) \geq (2)(1)$; (ii) no: $-2 \leq 1$ and $-1 \leq 1$ too, but $2 \geq 1$.

9. P7: Assume $x \leq y$ and $r \leq 0$. Let $r = -s$. Then by P6, $sx \leq sy$. By P5, $sx - sy \leq sy - sy = 0$. By P5, $-sx + sx - sy \leq -sx$. Hence $-sy \leq -sx$. So $ry \leq rx$.

    P8: If $z \geq 0$, let $x = 0$, $y = z$ and $r = z$ in P6. Then $0 \times z \leq z$. So $0 \leq z^2$. If $z < 0$, let $x = z$, $y = 0$ and $r = z$ in P7. Then $z \times 0 \leq z \times z$. So $0 \leq z^2$.

10. 78 kg.

11. She travelled 25 km in $3\frac{1}{2}$ hours. So her average speed was $7\frac{1}{7}$ kmh^{-1}.

12. If a project is worth three times as much as another it is relatively weighted by 3 against the other project. So the relative weights are the w_i and the scores on the projects are the x_i.

    Grade point averages are also determined by weighted averages.

    The slugging average (or percentage) of a batter in baseball is weighted by 1 if the batter gets a single hit, by 2 if he gets a double and so on.

Can you find a weighted mean in the working of the last exercise?

13. (a) Chester could have had figures of something like 100 runs for 2 wickets and 25 runs for 1 wicket to give him an average of $125/3 = 41.66$. Slim's figures could have been 51 for 1 and 151 for 6 for a season long average of $202/7 = 28.85$.

    (b) Chester would have won if he and Slim had both taken 1 wicket in the first half of the season and 6 in the second.

14. From the last exercise it can be seen that the last inequality will not always hold. Does it make any difference if we allow negative numbers?

15. Here $c = \frac{a+b}{2}$, so $c - a = \frac{b-a}{2}$ and $b - c = \frac{b-a}{2}$.

16. Since $\frac{\sqrt{a_1 a_2}}{a_1} = \frac{a_2}{\sqrt{a_1 a_2}} = \sqrt{\frac{a_2}{a_1}}$, there is a common ratio between consecutive terms.

    If $0 < a_1 < a_2$, then $\sqrt{a_1 a_2} < \sqrt{a_2^2} = a_2$ and $\sqrt{a_1 a_2} > \sqrt{a_1^2} = a_1$.

17. (i) $A(5) = 12.4$, $G(5) = 2^{15/5} = 8$;

    (ii) $A(4) = 30$, $G(4) = 3^{5/2} = 9\sqrt{3}$;

    (iii) $A(5) = 32.4$, $G(5) = (2^5 \times 5^5)^{1/5} = 2 \times 5 = 10$.

18. (i) $A(5) > G(5)$; (ii) $A(4) > G(4)$; (iii) $A(5) > G(5)$.

    The arithmetic mean is larger in each case.

19. Let $A(2) = a_1 + a_2$ and $G(2) = \sqrt{a_1 a_2}$.

    Now $A(2)^2 - G(2)^2 = \frac{1}{4}[(a_1 + a_2)^2 - 4a_1 a_2] = \frac{1}{4}(a_1 - a_2)^2$.

    This is non-negative for all a_1, a_2, so $A(2)^2 \geq G(2)^2$. Since $A(2) \geq 0$ and $G(2) \geq 0$, then $A(2) \geq G(2)$.

20. From the last exercise we see that if $A(2) = G(2)$, then $(a_1 - a_2)^2 = 0$. Hence $a_1 = a_2$. Clearly $A(2) = G(2)$ if and only if $a_1 = a_2 \geq 0$.

21. The conjecture is based on the results of the Exercises 17–19. You might also like to prove it for $n = 3$ before you go any further.

Conjecture. *For a_i non-negative, $i = 1, 2, \ldots, n$, $A(n) \geq G(n)$.*

Proof. We prove this conjecture using mathematical induction. Since we know that $A(2) \geq G(2)$ for a_1, $a_2 \geq 0$, I will assume that $A(k) \geq G(k)$.

Let $G = G(k+1) = (a_1 a_2 \cdots a_{k+1})^{\frac{1}{k+1}}$. Now, in the expression, E, below, I'll take G, $k - 1$ times. So let

$$E = (a_1 + a_2 + \cdots + a_k) + (a_{k+1} + G + G + \cdots + G)$$
$$= kA(k) + k\mathbb{A}(k),$$

where $\mathbb{A}(k)$ is an arithmetic mean in terms of a_{k+1} and G.

Hence by applying the inductive hypothesis $A(k) \geq G(k)$ twice, I get

$$E \geq k(a_1 a_2 \cdots a_k)^{\frac{1}{k}} + k(a_{k+1} G^{k-1})^{\frac{1}{k}}$$
$$= k[(a_1 a_2 \cdots a_k)^{\frac{1}{k}} + (a_{k+1} G^{k-1})^{\frac{1}{k}}]$$
$$\geq k\{2[(a_1 a_2 \cdots a_k)^{\frac{1}{k}} (a_{k+1} G^{k-1})^{\frac{1}{k}}]^{\frac{1}{2}}\}$$
$$\text{(since } A(2) \geq G(2))$$
$$= 2k(a_1 a_2 \cdots a_{k+1} G^{k-1})^{\frac{1}{2k}}$$
$$= 2k(G^{k+1} G^{k-1})^{\frac{1}{2k}}$$
$$= 2kG.$$

Hence $a_1 + a_2 + \cdots + a_{k+1} \geq (k+1)G$ and the result follows. □

I now make the conjecture for which we have almost no previous data, except for $n = 2$. However, it looks good if you try $n = 3$ and 4, say. It also looks as if you might be able to "bend" the proof above to get the result.

Conjecture. $A(n) = G(n)$ *if and only if all the a_i are equal.*

Proof. Using the inductive approach of the last exercise, we may assume that $A(k) = G(k)$ if and only if $a_1, a_2, \ldots, a_k$ are equal and non-negative. From the first inequality of the proof of $A(n) \geq G(n)$, we have equality precisely when $a_1 = a_2 = \cdots = a_k$ and $a_{k+1} = G$. Hence $a_{k+1} = a_1$. □

22. Since $A(2) \geq G(2)$, $a_1 + a_2 \geq 2\sqrt{a_1 a_2}$, with equality when $a_1 = a_2$.
    Hence $(a+b)(b+c)(c+a) \geq 2\sqrt{ab} \cdot 2\sqrt{bc} \cdot 2\sqrt{ca} = 8abc$. Equality is gained when $a = b = c$.
23. Since $A(3) \geq G(3)$, $a_1 + a_2 + a_3 \geq 3(a_1 a_2 a_3)^{\frac{1}{3}}$. Let $a_1 = x^3$, $a_2 = y^3$, $a_2 = z^3$ and the result follows.
24. Repeat the approach of the last exercise with $a_1 = x^3 y^3$, $a_2 = y^3 z^3$ and $a_3 = z^3 x^3$.
25. This will depend on what you try, but you might be able to generalise or extend some of the results of the previous exercises.
26. This problem requires the application of $A(3) \geq G(3)$ despite the fact that the left side has four terms. Applying $A(3) \geq G(3)$ to a, b, c gives

$$(a + b + c)^3 \geq 3^3 abc.$$

But $(s - d) = a + b + c$, so $(s - d)^3 \geq 3^3 abc$. Similarly we get $(s - a)^3 \geq 3^3 bcd$, $(s - b)^3 \geq 3^3 acd$, $(s - c)^3 > 3^3 abd$.

Multiplying the last four inequalities gives
$$(s-a)^3(s-b)^3(s-c)^3(s-d)^3 \geq 3^{12}a^3b^3c^3d^3.$$
Taking cube roots and noting that all factors are positive, gives
$$(s-a)(s-b)(s-c)(s-d) \geq 81abcd.$$
However, equality is only achieved in $A(3) \geq G(3)$ if $a_1 = a_2 = a_3$. Since not all of a, b, c, d are equal then not all of $s-a$, $s-b$, $s-c$, $s-d$ are equal. The strict inequality therefore follows.

27. Now $[(1+a_1)(1+a_2)\cdots(1+a_n)]^{\frac{1}{n}}$
$\leq [(1+a_1)+(1+a_2)+\cdots+(1+a_n)]/n.$

Hence $(1+a_1)(1+a_2)\cdots(1+a_n) \leq \left(\frac{n+s}{n}\right)^n = \left(1+\frac{s}{n}\right)^n = \sum_{k=0}^n \frac{\binom{n}{k}s^k}{n^k}$.

But $\frac{\binom{n}{k}s^k}{n^k} = \frac{n!s^k}{k!(n-k!)n^k} = \left[\frac{n(n-1)\cdots(n-k+1)}{n^k}\right]\frac{s^k}{k!} \leq \frac{s^k}{k!}$. Since the term in the square bracket is less than one. Hence the inequality follows.

28. Let $\tan\theta = a$. Since $A(2) \geq G(2)$, $a + 1/a \geq 2\sqrt{a \cdot \frac{1}{a}} = 2$. Hence for $\tan\theta > 0$, $\tan\theta + \cot\theta \geq 2$. (This is already more general than θ being acute.) If $\tan\theta < 0$, let $a = -\tan\theta$.

From the above argument $-\tan\theta - \cot\theta \geq 2$, so $\tan\theta + \cot\theta \leq -2$.

29. Let $a_i = i$ in the $A(n) \geq G(n)$ inequality.

Then $(1\cdot2\cdot3\cdots n)^{\frac{1}{n}} \leq (1+2+3+\cdots+n)/n = \frac{n(n+1)}{2n} = \frac{(n+1)}{2}$.

Hence $n! \leq \frac{(n+1)^n}{2}$. However, equality only holds in the arithmetic-geometric mean inequality, when all the a_i are equal. As this is not the case and $n > 1$, we have $n! < \frac{(n+1)^n}{2}$.

30. From $A(n) \geq G(n)$ we have
$$\frac{1}{n}\left[\frac{a_1}{a_2} + \frac{a_2}{a_3} + \cdots + \frac{a_n}{a_1}\right] \geq \left(\frac{a_1}{a_2} \cdot \frac{a_2}{a_3} \cdots \frac{a_n}{a_1}\right)^{\frac{1}{n}} = 1^{\frac{1}{n}} = 1.$$

Hence the inequality follows.

31. Since $A(2) \geq G(2)$, $a + bx^4 \geq 2\sqrt{abx^4}$. Hence $\frac{a+bx^4}{x^2} \geq 2\sqrt{ab}$. So the left side of the inequality is a minimum when we have equality. This occurs when $a + bx^4 = 2x^2\sqrt{ab}$. So $bx^4 - 2x^2\sqrt{ab} + a = 0$, which, by the quadratic formula, gives $x^2 = \sqrt{\frac{a}{b}}$.

The expression $\frac{a+bx^4}{x^2}$ is therefore minimised when $x = \pm(ab^{-1})^{\frac{1}{4}}$.

32. The trick here is to use the $A(3) \geq G(3)$ inequality to achieve a constant upper value for $x^2(9-x)$. This upper bound is achieved when the components of the inequality are equal.

So use the fact that
$$\frac{x}{2} \cdot \frac{x}{2}(9-x) \leq \left(\frac{\frac{x}{2}+\frac{x}{2}+9-x}{3}\right)^3 = 3^3.$$

This maximum is achieved when $\frac{x}{2} = \frac{x}{2} = 9-x$. That is, when $x = 6$.

33. Using the idea of the last exercise we have to have $\frac{5+2x}{a}$ and $\frac{(3-x)}{b}$ so that when we add $\frac{3(5+2x)}{a}$ to $\frac{5(3-x)}{b}$ we can cancel the x terms. This can be done by choosing $a = 6$ and $b = 5$. Hence

$$\left(\frac{5+2x}{6}\right)^3 \left(\frac{3-x}{5}\right)^5 \leq \left[\frac{3\left(\frac{5+2x}{6}\right) + 5\left(\frac{3-x}{5}\right)}{8}\right]^8 = \left(\frac{11}{16}\right)^8.$$

So $(5+2x)^3(3-x)^5$ has a maximum of $\frac{6^3 5^3 11^8}{16^8} = \left(\frac{33}{8}\right)^3 \left(\frac{55}{16}\right)^5$. This maximum occurs when $x = -\frac{7}{16}$.

34. In the $A(n) \geq G(n)$ inequality, let $a_1 = a$ and $a_i = b$ for $i = 2, 3, \ldots, n+1$. Then we get $(ab^n)^{\frac{1}{(n+1)}} \leq \frac{a+b+\cdots+b}{n+1} = \frac{a+nb}{n+1}$.

35. As in earlier problems, the trick is to take $\left(\frac{x}{a}\right)^2 \left(\frac{y}{b}\right)^3 z$ so that

$$2\left(\frac{x}{a}\right) + 3\left(\frac{y}{b}\right) + z = x + y + z = c.$$

So choose $a = 2$ and $b = 3$.

Now $\left(\frac{x}{a}\right)^2 \left(\frac{y}{b}\right)^3 z \leq \left[\frac{x+y+z}{6}\right]^6$ by $G(3) \leq A(3)$.

Hence $x^2 y^3 z \leq \frac{2^2 3^3 c^6}{6^6} = \frac{c^6}{432}$.

36. This is the last exercise, but in reverse. Since $\left[\frac{x+y+z}{6}\right]^6 \geq \left(\frac{x}{2}\right)^2 \left(\frac{y}{3}\right)^3 z$ and $x^2 y^3 z = 108$, then $x + y + z \geq 6\left[\frac{106}{2^2 3^3}\right]^{\frac{1}{6}} = 6$. The minimum is obtained when $\frac{x}{2} = \frac{y}{3} = z$. Substituting for z in $x^2 y^3 z = 108$ gives $(2z)^2(3z)^3 z = 108$. Thus $z = 1$, $x = 2$ and $y = 3$.

37. We need to choose a and b so that $3\left(\frac{x}{a}\right) + 2\left(\frac{y}{b}\right) + z = 6x + 4y + z$. Take $a = b = \frac{1}{2}$. Now $(2x)^3 (2y)^2 z \leq \left[\frac{6x+4y+z}{6}\right]^6 = 6^{-6}$.

So $x^3 y^2 z \leq 6^{-6} z^{-5}$. This maximum occurs when $2x = 2y = z$ and $6x + 4y + z = 1$. So $3z + 2z + z = 1$ gives $z = 1/6$, $x = 1/12$, $y = 1/12$.

38. Let the dimensions of the rectangle be a and b. If the area is A, then $A = ab \leq \left(\frac{a+b}{2}\right)^2 \leq \frac{P^2}{16}$. The area of a rectangle with perimeter P is therefore at most $\frac{P^2}{16}$. The maximum is achieved when $a = b$. The rectangle is therefore a square.

39. For a rectangle with perimeter P and fixed area A,

$$P = 2(a + b) \geq 4\sqrt{ab} = 4\sqrt{A}.$$

So a rectangle with area A has perimeter at least $4\sqrt{A}$. This minimum value is obtained when $a = b$, that is, for the square.

40. Let the lengths of the sides of the box be a, b, c such that $x = ab$, $y = bc$, $z = ca$. Then $V = abc \leq \sqrt{xyz}$ and $A = 2(x + y + z)$.

Now $\sqrt[3]{xyz} \leq \frac{1}{3}(x + y + z) = \frac{A}{6}$. So $V \leq \left(\frac{A}{6}\right)^{\frac{3}{2}}$. Hence the volume of any box is at most $\left(\frac{A}{6}\right)^{\frac{3}{2}}$ and this maximum value is obtained when $x = y = z$, that is for the cube.

41. By Heron's formula for any triangle, the area A, is given by $A = \sqrt{s(s-a)(s-b)(s-b)(s-c)}$, where s is the semi-perimeter. Since $A(3) \geq G(3)$, $(s-a)(s-b)(s-c) \leq \left[\frac{(s-a)+(s-b)+(s-c)}{3}\right]^3 = \left(\frac{s}{3}\right)^3 = \left(\frac{P}{6}\right)^3$. Hence $A \leq \sqrt{s}\left(\frac{P}{6}\right)^{\frac{3}{2}} = \left(\frac{P}{2}\right)^{\frac{1}{2}}\left(\frac{P}{6}\right)^{\frac{3}{2}} = \frac{P^2}{12\sqrt{3}}$.

For all triangles then, the area is at most $\frac{P^2}{12\sqrt{3}}$. This maximum is reached when $s - a = s - b = s - c$. That is, when $a = b = c$ and we have an equilateral triangle.

42. For $a_i = a$, $i = 1, 2, \ldots, n$,

$$A(n) = \left(\frac{a + a + \cdots + a}{n}\right) = a;$$

$$G(n) = \sqrt[n]{aa \ldots a} = a;$$

$$H(n) = \frac{n}{\frac{1}{a} + \frac{1}{a} + \cdots + \frac{1}{a}} = a.$$

43. Since $\frac{1}{a} - \frac{1}{b} = \frac{1}{b} - \frac{1}{c}$, then $\frac{2}{b} = \frac{1}{a} + \frac{1}{c}$. Hence $\frac{1}{b} = \frac{1}{2}\left(\frac{1}{a} + \frac{1}{c}\right)$ and $b = \frac{2}{\frac{1}{a} + \frac{1}{c}}$ as required.

Now look back at Exercises 15 and 16.

44. (i) $\frac{160}{31}$; (ii) $\frac{81}{100}$; (iii) $\frac{5000}{1173}$.

45. $G(2)^2 - H(2)^2 = a_1 a_2 - \frac{4a_1^2 a_2^2}{(a_1 + a_2)^2} = \frac{a_1 a_2 (a_1 - a_2)^2}{(a_1 + a_2)^2} \geq 0$. Since $G(2) \geq 0$, and $G(2) - H(2) \geq 0$, $G(2) \geq H(2)$. Clearly $G(2) = H(2)$ when $a_1 = a_2$.

46. In $G(3)$, let $\alpha_1 = \frac{1}{\alpha_1}$, $a_2 = \frac{1}{\alpha_2}$, $a_3 = \frac{1}{\alpha_3}$, where $a_i > 0$. Since $A(3) \geq G(3)$, we get $\frac{1}{3}\left(\frac{1}{\alpha_1} + \frac{1}{\alpha_2} + \frac{1}{\alpha_3}\right) \geq \left[\frac{1}{\alpha_1} + \frac{1}{\alpha_2} + \frac{1}{\alpha_3}\right]^{\frac{1}{3}}$. Hence, since both sides are positive, $(\alpha_1 \alpha_2 \alpha_3)^{\frac{1}{3}} \geq \frac{3}{\frac{1}{\alpha_1} + \frac{1}{\alpha_2} + \frac{1}{\alpha_3}}$. So we have proved $G(3) \geq H(3)$ provided $a_i > 0$ and hence $\alpha_i > 0$.

Equality holds here when $A(3) = G(3)$. That is when $\frac{1}{\alpha_1} = \frac{1}{\alpha_2} = \frac{1}{\alpha_3}$ or equivalently $\alpha_1 = \alpha_2 = \alpha_3$.

47. Using $a_i = \frac{1}{\alpha_i}$ in the $A(n) \geq G(n)$ inequality, the fact that $G(n) \geq H(n)$ follows using an argument similar to that of the last exercise. Again equality holds when each a_i term of the mean is the same.

48. Since $A(3) \geq H(3)$, $\frac{1}{3}[(a+b) + (b+c) + (c+a)] \geq \frac{3}{\frac{1}{a+b} + \frac{1}{b+c} + \frac{1}{c+a}}$. Hence $2(a + b + c)\left[\frac{1}{a+b} + \frac{1}{b+c} + \frac{1}{c+a}\right] \geq 9$ and the required inequality follows.

49. From the last exercise, $\frac{a+b+c}{a+b} + \frac{a+b+c}{b+c} + \frac{a+b+c}{c+a} \geq \frac{9}{2}$.

So $1 + \frac{c}{a+b} + 1 + \frac{a}{b+c} + 1 + \frac{b}{c+a} \geq \frac{9}{2}$.

The required inequality follows since $\frac{9}{2} - 3 = \frac{3}{2}$. Equality holds when $a = b = c$.

50. For $H(2)$ to equal $A(2)$ it is necessary and sufficient for a_1 to equal a_2 and be non-zero.

    Necessity. For $H(2)$ to be defined a_1, $a_2 \neq 0$, and $a_1 + a_2 \neq 0$. If $H(2) = A(2)$, then $\frac{2a_1a_2}{a_1+a_2} = \frac{a_1+a_2}{2}$. Hence $4a_1a_2 = (a_1 + a_2)^2$. So $(a_1 - a_2)^2 = 0$. This gives $a_1 = a_2$.

    Sufficiency. If $a_1 = a_2 \neq 0$, then $H(2) = a_1$ and $A(2) = a_1$. Hence $H(2) = A(2)$.

51. Now $H(2) = \frac{2a_1a_2}{a_1+a_2}$ and $A(2) = \frac{a_1+a_2}{2}$. Hence their geometric mean is
    $$\sqrt{\left(\frac{2a_1a_2}{a_1+a_2}\right)\left(\frac{a_1+a_2}{2}\right)} = \sqrt{a_1a_2} \text{ as required.}$$

52. $Q(n) = [(a^2 + a^2 + \cdots + a^2)/n]^{\frac{1}{2}} = \sqrt{\frac{na^2}{n}} = a$.

53. Now $Q(n)^2 - A(n)^2 = \frac{a_1^2+a_2^2}{2} - \left(\frac{a_1+a_2}{2}\right)^2 = \frac{1}{4}(a_1 - a_2)^2 \geq 0$. Hence, since $Q(2) \geq 0$, it follows that $Q(2) \geq A(2)$. Furthermore, $Q(2) = A(2)$ if and only if $a_1 = a_2$. So $Q(2) \geq A(2) \geq G(2) \geq H(2)$ with equality precisely when $a_1 = a_2$.

    In general $Q(n) \geq A(n)$ with equality when the a_i are all equal. This follows since $Q(n)^2 - A(n)^2 = \frac{1}{n}\sum_{i=1}^{n}(a_i - a_{i+1})^2$, where $a_{n+1} = a_1$.

    Actually $Q(n)^2 - A(n)^2 = \frac{1}{n}\sum_{i=1}^{n}(a_i - \bar{a})^2$, where $\bar{a} = A(n)$. Does this remind you of any statistics?

54. Now $\frac{a+b}{2} \leq \sqrt{\frac{a^2+b^2}{2}}$ since $A(2) \leq Q(2)$. Hence
    $$a + b \leq \sqrt{2}\sqrt{a^2 + b^2} = c\sqrt{2}.$$

55. Let $E = \left\{\frac{1}{2}\left[\left(a + \frac{1}{a}\right)^2 + \left(b + \frac{1}{b}\right)^2\right]\right\}^{\frac{1}{2}}$. Since $Q(2) \geq A(2)$ and $a + b = 1$, then $E \geq \frac{1}{2}\left[\left(a + \frac{1}{a}\right) + \left(b + \frac{1}{b}\right)\right] = \frac{1}{2}\left[1 + \frac{1}{ab}\right]$.

    But $G(2) \leq A(2)$, $ab \leq \left(\frac{a+b}{2}\right)^2 = \frac{1}{4}$. Hence $\frac{1}{ab} \geq 4$ and $E \geq \frac{5}{2}$. So $E^2 \geq \frac{25}{4}$ and the result follows.

56. Now $\frac{x+y}{2} \leq \sqrt{\frac{x^2+y^2}{2}}$ since $A(2) \leq Q(2)$. This means that
    $$(x + y)^2 \leq 2(x^2 + y^2).$$
    Similarly $(y + z)^2 \leq 2(y^2 + z^2)$ and $(z + x)^2 \leq 2(z^2 + x^2)$. Adding the last three inequalities gives the required result.

57. We use the $A(n) \leq Q(n)$ inequality with $n = 12a + 13b + 4c$.
    $$f(x, y, z) = 12a\left(\frac{x}{a}\right) + 3b\left(\frac{xy}{b}\right) + 4c\left(\frac{z}{c}\right)$$
    $$\leq (12a + 3b + 4c)\left\{\frac{12a\left(\frac{x}{a}\right)^2 + 3b\left(\frac{xy}{b}\right)^2 + 4c\left(\frac{z}{c}\right)^2}{12a + 3b + 4c}\right\}^{\frac{1}{2}}.$$
    To get $12a\left(\frac{x}{a}\right)^2 + 3b\left(\frac{xy}{b}\right)^2 + 4c\left(\frac{z}{c}\right)^2 = x^2 + y^2 + z^2$ we require $a = 12$, $b = 3$ and $c = 4$. Hence $f(x, y, z) = 169\left[\frac{x^2+y^2+z^2}{169}\right]^{\frac{1}{2}} = 13(x^2 + y^2 + z^2)^{\frac{1}{2}} = 13$.

The maximum value of 13 is attained when $\frac{x}{12} = \frac{y}{3} = \frac{z}{4}$. Substituting in $x^2 + y^2 + z^2 = 1$ gives $9z^2 + \frac{9z^2}{16} + z^2 = 1$. Hence $z = \pm\frac{4}{13}$, $x = \pm\frac{12}{13}$, $y = \pm\frac{3}{13}$. The maximum is reached at the point $\left(\frac{12}{13}, \frac{3}{13}, \frac{4}{13}\right)$ because the point $\left(-\frac{12}{13}, -\frac{3}{13}, -\frac{4}{13}\right)$ leads to a negative value of f.

58. (i) no, because when $x = 1$ or $x = -1$ we get $y = 1$, so two values of x go to one value of y. We can make this a one-to-one function by limiting the domain to a subset of the non-negative real numbers or a subset of the non-positive real numbers; (ii) yes (only one value of x will give any particular value of y); (iii) yes; (iv) no but choose any subset of $\left(-\frac{\pi}{2}, \frac{\pi}{2}\right)$, for example, as the domain and you will have a one-to-one function on that domain.

59. (i) the quadratic mean; (ii) the harmonic mean.

60. Let $f(x) = \log_e x$. Then $f^{-1}(x) = e^x$. So here

$$M_f = e^{\frac{1}{n}(\log a_1 + \log a_2 + \cdots + \log a_n)} = e^{\log(a_1 a_2 \cdots a_n)^{\frac{1}{n}}}$$
$$= (a_1 a_2 \cdots a_n)^{\frac{1}{n}}.$$

This is the geometric mean. (Or see Exercise 62(ii).)

61. This is up to you. Did you come across anything that follows in this chapter or anything that you think is interesting?

62. (i) If $p > q$, then $M_p \geq M_p$; (ii) $M_0 = G(n)$.

63. (a) Let $a_1 = 1$, $a_2 = 3$, $b_1 = 3$, $b_2 = 1$. Then $\overline{A} = 2$, $\overline{B} = 2$ and $\overline{C} = 3$. So $\overline{A} \times \overline{B} \neq \overline{C}$.

(b) $\overline{A} = \frac{1}{2}(a_1 + a_2)$, $\overline{B} = \frac{1}{2}(b_1 + b_2)$, $\overline{C} = \frac{1}{2}(a_1 b_1 + a_1 b_2)$. Then $\overline{C} - \overline{A} \times \overline{B} = \frac{1}{2}(a_1 b_1 + a_2 b_2) - \frac{1}{4}(a_1 + a_2)(b_1 + b_2)$. Regular algebra reduces this to $\frac{1}{4}(a_1 - a_2)(b_1 - b_2)$. This term is non-zero since $a_1 \geq a_2$ and $b_1 \geq b_2$. Hence $\overline{C} \geq \overline{A} \times \overline{B}$.

(c) To do this part put $n = 3$ in (d) below. However, it would help a lot if you did the case $n = 3$ first as the sigma notation I use in (d) may be a bit complicated if this is the first time you are looking at such arguments. As you do (d) you can check the steps against the case $n = 3$.

(d) Now

$$n^2(\overline{C} - \overline{A} \times \overline{B}) = n \sum_{i=1}^{n} a_i b_i - \sum_{j=1}^{n} a_i \sum_{i=1}^{n} b_j$$
$$= \frac{1}{2}\left\{ \sum_{i=1}^{n} \sum_{i=1}^{n}(a_i b_i - a_i b_j) \right.$$
$$\left. + \sum_{j=1}^{n} \sum_{j=1}^{n}(a_j b_j - a_j b_i) \right\}^*$$
$$= \frac{1}{2}\left\{ \sum_{i=1}^{n} \sum_{j=1}^{n}(a_i b_i - a_i b_j + a_j b_j - a_j b_i) \right\}$$
$$= \frac{1}{2} \sum_{i=1}^{n} \sum_{j=1}^{n}(a_i - a_j)(b_i - b_j).$$

Since this last expression is positive or zero (the a_i and b_j are non-decreasing), the inequality holds.

(Step * needs a little justification but it should not be too hard.)

(e) If A and B are non-increasing sequences, then $a_i - a_j$ and $b_i - b_j$ always have the same sign. So the last expression in (d) is still positive or zero and $\overline{C} \geq \overline{A} \times \overline{B}$.

64. For x, $y \geq 1$, the terms $1, x, x^2, \ldots, x^n$ and $1, y, y^2, \ldots, y^n$ are non-decreasing. Hence by Chebycheff, the inequality holds.

If $0 < x$, $y \leq 1$, then $1, x, x^2, \ldots, x^n$ and $1, y, y^2, \ldots, y^n$ are non-increasing. Hence by Chebycheff the same inequality holds as for x, $y \geq 1$.

If $0 < x < 1 < y$, then $1, x, x^2, \ldots, x^n$ is non-increasing and $1, y, y^2, \ldots, y^n$ is non-decreasing. If we return to the solution of the last exercise, we see that for a_i non-increasing and b_i non-decreasing, $(a_i - a_j)$ and $(b_i - b_j)$ have opposite signs. Hence $\overline{C} \leq \overline{A} \times \overline{B}$. So

$$\sum_{k=0}^{n} x^k \sum_{k=0}^{n} y^k \geq (n+1) \sum_{k=0}^{n} x^k y^k.$$

65. Equality holds if either sequence is constant. But is this a necessary and sufficient condition?

66. RHS $-$ LHS $= (a_1^2 + a_2^2)(b_1^2 + b_2^2) - (a_1 b_1 + a_2 b_2)^2$
$= a_1^2 b_1^2 + a_1^2 b_2^2 + a_2^2 b_1^2 + a_2^2 b_2^2 - (a_1^2 b_1^2 + 2a_1 b_1 a_2 b_2 + a_2^2 b_2^2)$
$= a_1^2 b_2^2 + a_2^2 b_1^2 - 2a_1 b_1 a_2 b_2 = (a_1 b_2 - a_2 b_1)^2.$

Any square is positive or zero, so RHS $-$ LHS ≥ 0. Hence the inequality holds.

67. Similar reasoning to the last exercise leads to the difference between the two sides of the inequality being

$$(a_1 b_2 - a_2 b_1)^2 + (a_1 b_3 - a_3 b_1)^2 + (a_2 b_3 - a_3 b_2)^2.$$

Since this is the sum of three squares, the inequality holds.

The generalisation of this and the previous exercise is obvious. However, it looks as if it might be a pain to prove.

68. A generalisation might be $\left(\sum_{i=1}^{n} a_i b_i \right)^2 \leq \sum_{i=1}^{n} a_i^2 \sum_{i=1}^{n} b_i^2$. You might attempt to prove this by induction.

69. Clearly if $a_i = b_i$ for $i = 1, 2, \ldots, n$ then equality holds. But is this the only possibility?

You will know from the work you did in the last exercise that

$$\sum_{1 < i < j < n} (a_i b_j - a_j b_i)^2 = \sum_{i=1}^{n} a_i^2 \sum_{i=1}^{n} b_i^2 - \left(\sum_{i=1}^{n} a_i b_i \right)^2.$$

So equality holds in the C-S-B inequality if $a_i b_j - a_j b_i = 0$, for all i, $j = 1, 2, \ldots, n$ with $i < j$. This is clearly so if all the a_i's are zero and all the b_i's are zero.

So suppose that not all the b_i are zero. Without loss of generality we may assume that $b_n \neq 0$. (If not, one of the b_i's is non-zero, so rearrange them until $b_n \neq 0$.) Since $a_i b_j - a_j b_i = 0$ for $i, j = 1, 2, \ldots, n$ with $i < j$, then $a_i b_n - a_n b_i = 0$ for $i = 1, 2, \ldots, n - 1$. Hence $a_i = \frac{a_n}{b_n} b_i$.

Let $\frac{a_n}{b_n} = \lambda$, where λ is some real number. So far we have shown that $a_i = \lambda b_i$ for $i = 1, 2, \ldots, n - 1$. Obviously $a_n = \lambda b_n$. Hence $a_i = \lambda b_i$ for $i = 1, 2, \ldots, n$ and λ some real number.

By the symmetry of the inequality we could also have $b_i = \mu a_i$ for $i = 1, 2, \ldots, n$ and μ real.

(Note that, if $\lambda = 0$, this covers the case where all the a_i are zero and if $\mu = 0$, this covers the case where all the b_i are zero.)

So if the C-S-B inequality is an equality, $a_i = \lambda b_i$ for $i = 1, 2, \ldots, n$ for some real number λ or $b_i = \mu a_i$ for $i = 1, 2, \ldots, n$ and some real number μ.

Suppose $a_i = \lambda b_i$ for $i = 1, 2, \ldots, n$ and $\lambda \in \mathbb{R}$. Then

$$\sum_{i=1}^{n} a_i^2 \sum_{i=1}^{n} b_i^2 = \sum_{i=1}^{n} \lambda^2 b_i^2 \sum_{i=1}^{n} b_i^2$$
$$= \left(\sum_{i=1}^{n} \lambda^2 b_i^2 \right)^2 = \left(\sum_{i=1}^{n} a_i b_i \right)^2.$$

In other words, if $a_i = \lambda b_i$, then the C-S-B inequality is an equality. Similarly we get equality if $b_i = \mu a_i$.

We have therefore shown that the necessary and sufficient conditions for the Cauchy–Schwarz inequality to be an equality is that $a_i = \lambda b_i$ for $i = 1, 2, \ldots, n$ with some fixed real number λ or $b_i = \mu a_i$ for some real μ.

70. To use the C-S-B inequality, I need to find a_i and b_i. Since I want to maximise f, I require $\sum_{i=1}^{3} a_i b_i = x + 2y + 4z$. This can be done by taking various values for a_i and b_i. However, I also require something on the right-hand side to involve $x^2 + 2y^2 + z^2$.

Hence I'll choose $a_1 = x$, $a_2 = y\sqrt{2}$, $a_3 = z$, $b_1 = 1$, $b_2 = \sqrt{2}$, $b_3 = 4$. Then $\sum_{i=1}^{3} a_i b_i = x + 2y + 4z$, $\sum_{i=1}^{3} a_i^2 = x^2 + 2y^2 + z^2$, and $\sum_{i=1}^{3} b_i^2 = 19$.

So $x + 2y + 4z \leq (x^2 + 2y^2 + z^2)^{\frac{1}{2}} \sqrt{19} = \sqrt{19}$. The maximum value of f is therefore 19.

To find where the maximum occurs, note that equality is obtained in the C-S-B inequality when $a_i = \lambda b_i$ (or $b_i = \mu a_i$). Since for $x + 2y + 4z$

to equal $\sqrt{19}$, not all of the a_i or b_i are zero, we may assume $a_i = \lambda b_i$. Hence $x = \lambda$, $y = \lambda$, $z = 4\lambda$. So $\lambda + 2\lambda + 16\lambda = \sqrt{19}$. This gives $\lambda = \frac{1}{\sqrt{19}}$. The maximum value of f therefore, comes at the point $\left(\frac{1}{\sqrt{19}}, \frac{1}{\sqrt{19}}, \frac{4}{\sqrt{19}}\right)$.

71. Using the method of the last exercise but realising that we require a **minimum** value for f, choose $a_1 = x$, $a_2 = y$, $a_3 = z$, $b_1 = 1$, $b_2 = 1$, $b_3 = \sqrt{2}$.

Hence $(x+y+z)^2 \le (x^2 + y^2 + z^2/2)(1+1+2)$. So $x^2 + y^2 + z^2/2 \ge \frac{25}{2}$.

We get equality when $x = \lambda$, $y = \lambda$, $\frac{z}{\sqrt{2}} = \lambda\sqrt{2}$. Substituting in $x^2 + y^2 + z^2/2 = 25$ gives $\lambda = \frac{5}{2}$. So the minimum is achieved at $\left(\frac{5}{2}, \frac{5}{2}, 5\right)$.

72. Let $a_1 = x$, $a_2 = y$, $a_3 = z$, $b_1 = 12$, $b_2 = 3$, $b_3 = 4$.

73. $p = q = 2$.

74. Suppose that $\lambda a_i^p = \mu b_i^p$. Then
$$\sum\nolimits_{i=1}^{n} a_i b_i = \left(\sum\nolimits_{i=1}^{n} \left(\frac{\mu}{\lambda} b_i^q\right)^{\frac{1}{p}} b_i\right) = \left(\frac{\mu}{\lambda}\right)^{\frac{1}{p}} \sum\nolimits_{i=1}^{n} b_i^q.$$
Since $\frac{1}{p} + \frac{1}{q} = 1$ implies that $\frac{p+q}{p} = q$.

Further,
$$\left(\sum\nolimits_{i=1}^{n} a_i^p\right)^{\frac{1}{p}} \left(\sum\nolimits_{i=1}^{n} b_i^q\right)^{\frac{1}{q}} = \left(\sum\nolimits_{i=1}^{n} \left(\frac{\mu}{\lambda} b_i^q\right)\right)^{\frac{1}{p}} \left(\sum\nolimits_{i=1}^{n} b_i^q\right)^{\frac{1}{q}}$$
$$= \left(\frac{\mu}{\lambda}\right)^{\frac{1}{p}} \left(\sum\nolimits_{i=1}^{n} b_i^q\right)^{\frac{1}{p}} \left(\sum\nolimits_{i=1}^{n} b_i^q\right)^{\frac{1}{q}} = \left(\frac{\mu}{\lambda}\right)^{\frac{1}{p}} \left(\sum\nolimits_{i=1}^{n} b_i^q\right),$$
since $\frac{1}{p} + \frac{1}{q} = 1$.

Actually the condition here is a necessary and sufficient condition for equality.

75. If $A + B = (c_1, c_2, \ldots, c_n)$, where $c_i = a_i + b_i$, then
$$|A + B| = \left(\sum\nolimits_{i=1}^{n} c_i^2\right)^{\frac{1}{2}} = \left(\sum\nolimits_{i=1}^{n} (a_i + b_i)^2\right)^{\frac{1}{2}}.$$
This is less than or equal to $|A| + |B|$ by the triangle inequality.

[The reason for doing this is that it gives us a way of adding two points (or vectors) in n-dimensional space. $|A|$ is the distance of the point A from the origin.]

76. If $A \cdot B = c_i$, where $c_i = a_i b_i$, then $|\sum_{i=1}^{n} c_i| = |A \cdot B|$ is less than or equal to $|A| \cdot |B|$ by the C-S-B inequality.

[The reason for defining $A \cdot B$ in this way is that it gives a way of multiplying points (or vectors) in n-dimensions. This product has many uses outside the C-S-B inequality.]

77. Equality holds if and only if, for some $\lambda, \mu \ge 0$, $a_i = \lambda b_i$ or $b_i = \mu a_i$, $1, 2, \ldots, n$.

Suppose $a_i = \lambda b_i$ with $\lambda \geq 0$. Then

$$\left(\sum_{i=1}^{n} (a_i + b_i)^2\right)^{\frac{1}{2}} = \left(\sum_{i=1}^{n} (b_i)^2 (\lambda + 1)^2\right)^{\frac{1}{2}}$$

$$= (\lambda + 1)\left(\sum_{i=1}^{n} (b_i)^2\right)^{\frac{1}{2}}$$

$$= \lambda \left(\sum_{i=1}^{n} (b_i)^2\right)^{\frac{1}{2}} + \left(\sum_{i=1}^{n} (b_i)^2\right)^{\frac{1}{2}}$$

$$= \left(\sum_{i=1}^{n} \lambda^2 b_i^2\right)^{\frac{1}{2}} + \left(\sum_{i=1}^{n} (b_i)^2\right)^{\frac{1}{2}}$$

$$= \left(\sum_{i=1}^{n} (a_i)^2\right)^{\frac{1}{2}} + \left(\sum_{i=1}^{n} (b_i)^2\right)^{\frac{1}{2}}.$$

(Where did we need $\lambda \geq 0$?)

Recall that, by the C-S-B inequality, $\sum_{i=1}^{n} a_i^2 \sum_{i=1}^{n} b_i^2 = \left(\sum_{i=1}^{n} a_i b_i\right)^2$ if and only if $a_i = \lambda b_i$, $\lambda \in \mathbb{R}$ and $i = 1, 2, \ldots, n$, or $b_i = \mu a_i$, $\mu \in \mathbb{R}$ and $i = 1, 2, \ldots, n$.

So now suppose the triangle inequality **is** an equality. Squaring both sides gives

$$\sum_{i=1}^{n} (a_i + b_i)^2$$

$$= \sum_{i=1}^{n} (a_i)^2 + 2\left(\sum_{i=1}^{n} (a_i)^2\right)^{\frac{1}{2}} \left(\sum_{i=1}^{n} (b_i)^2\right)^{\frac{1}{2}} + \sum_{i=1}^{n} (b_i)^2.$$

But $\sum_{i=1}^{n} (a_i + b_i)^2 = \sum_{i=1}^{n} (a_i)^2 + 2\left(\sum_{i=1}^{n} a_i b_i\right) + \sum_{i=1}^{n} (b_i)^2$.

Hence $\sum_{i=1}^{n} a_i b_i = \left(\sum_{i=1}^{n} (a_i)^2\right)^{\frac{1}{2}} \left(\sum_{i=1}^{n} (b_i)^2\right)^{\frac{1}{2}}$.

Now the C-S-B inequality is an equality if and only if $a_i = \lambda b_i$ or $b_i = \mu a_i$ for $\lambda, \mu \in R$. But if $\lambda < 0$, the last equality is not an equality.

78. Now

$$\sum_{i=1}^{n} (a_i + b_i)^2 = \sum_{i=1}^{n} (a_i^2 + 2a_i b_i + b_i^2)$$

$$= \sum_{i=1}^{n} a_i^2 + 2\sum_{i=1}^{n} a_i b_i + \sum_{i=1}^{n} b_i^2$$

$$\leq \sum_{i=1}^{n} a_i^2 + 2\left(\sum_{i=1}^{n} a_i^2\right)^{\frac{1}{2}} \left(\sum_{i=1}^{n} b_i^2\right)^{\frac{1}{2}} + \sum_{i=1}^{n} b_i^2$$

$$= \left[\left(\sum_{i=1}^{n} a_i^2\right)^{\frac{1}{2}} + \left(\sum_{i=1}^{n} b_i^2\right)^{\frac{1}{2}}\right]^2.$$

The inequality comes from an application of the C-S-B inequality.

Hence $\sum_{i=1}^{n} (a_i + b_i)^2 \leq \left[\left(\sum_{i=1}^{n} a_i^2\right)^{\frac{1}{2}} + \left(\sum_{i=1}^{n} b_i^2\right)^{\frac{1}{2}}\right]^2$.

So $\left(\sum_{i=1}^{n} (a_i + b_i)^2\right)^{\frac{1}{2}} \leq \left(\sum_{i=1}^{n} a_i^2\right)^{\frac{1}{2}} + \left(\sum_{i=1}^{n} b_i^2\right)^{\frac{1}{2}}$.

This is the triangle inequality.

79. $p = 2$.

80. Substituting xy for z^2 in (2) gives
$$x^2 + xy + y^2 = b^2. \tag{i}$$
Now from (1), $(x+y) - a = -z$, so $(x+y)^2 - 2a(x+y) + a^2 = z^2$. Hence from (3)
$$x^2 + xy + y^2 - 2a(x+y) = -a^2. \tag{ii}$$
Subtracting (ii) from (i) gives
$$x + y = \frac{a^2 + b^2}{2a}. \tag{iii}$$

Hence $z = \frac{a^2 - b^2}{2a}$, and so
$$xy = \frac{(a^2 - b^2)^2}{4a^2}. \tag{iv}$$
The variables x and y are therefore the roots of the quadratic equation
$$w^2 - \left(\frac{a^2 + b^2}{2a}\right) w + \frac{(a^2 - b^2)^2}{4a^2} = 0.$$
The discriminant of this equation is
$$\Delta = \frac{1}{4a^2}(3a^2 - b^2)(3b^2 - a^2).$$

Since x and y are real and distinct, $3a^2 \geq b^2$ and $3b^2 \geq a^2$ or $3a^2 \leq b^2$ and $3b^2 \leq a^2$. From (1) and (2) this latter condition cannot hold.

Further $\sqrt{\Delta} \leq \{\pm \frac{1}{2}\left[\frac{3a^2 - b^2}{2a} + \frac{3b^2 - a^2}{2a}\right]\}$ by the $A(2) \geq G(2)$ inequality. Hence $\sqrt{\Delta} \leq \pm\left(\frac{a^2 + b^2}{4a}\right)$, with the positive sign if $a > 0$ and the negative sign if $a < 0$. Equality holds when $a^2 = b^2$ which implies $z = 0$. Hence strict inequality holds.

Now solving the quadratic in w for x and y gives $x = \frac{a^2 + b^2}{4a} \pm \frac{1}{2}\sqrt{\Delta}$ and $y = \frac{a^2 + b^2}{4a} \mp \frac{1}{2}\sqrt{\Delta}$. Since $\sqrt{\Delta} \leq \left|\frac{a^2 + b^2}{4a}\right|$, and both the values for x and y are positive, then $a > 0$. So if $a > 0$ and $3a^2 > b^2 > \frac{1}{3}a^2$ then we have two values for x and y and $z = \frac{a^2 - b^2}{2a}$. Since $x \neq y$ and z is their geometric mean (by (3)), then x, y and z are all distinct positive numbers.

81. Let $P = a + b + c$. Now by Exercise 41, the equilateral triangle has largest area among all triangles with a given perimeter.
Hence $T \leq \frac{P^2\sqrt{3}}{36}$.
Now $P^2 + (a-b)^2 + (b-c)^2 + (c-a)^2 = 3(a^2 + b^2 + c^2)$. This gives $P^2 \leq 3(a^2 + b^2 + c^2)$, with equality if and only if $a = b = c$. So $T \leq \frac{(a^2 + b^2 + c^2)\sqrt{3}}{12}$. Thus $a^2 + b^2 + c^2 \geq 4T\sqrt{3}$.
Now the inequality is an equality if and only if $a = b = c$, in other words, when the triangle is equilateral.

82. Let the points I, I_1 and I_1', be the centres of the incircles and the excircle of the triangles ABC and ABD as shown in the following figure.

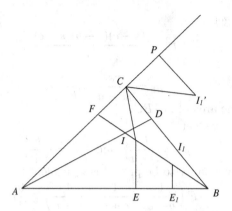

Let E, E_1 be on AB and F, P be on AC so that $IE \perp AB$, $I_1 E_1 \perp AB$, $IF \perp AC$, and $I_1' P \perp AC$. Then from the similar triangles $I_1' CP$ and ICF we have

$$\frac{\rho_1}{CP} = \frac{I_1' P}{CP} = \frac{CF}{IF} = \frac{s-c}{r}, \tag{1}$$

where a, b, c, s are the sides and semi-perimeter of $\triangle ABC$. It is well-known that

$$CP = AP - AC = \frac{b + AD + CD}{2} - b = \frac{AD + CD - b}{2}. \tag{2}$$

From (1) and (2) we get

$$\frac{r\rho_1}{s-c} = \frac{AD + CD - b}{2}. \tag{3}$$

Further, from the similarity of triangles BIE and $BI_1 E_1$ we obtain

$$\frac{\rho_1}{r} = \frac{I_1 E_1}{IE} = \frac{BE_1}{BE} = \frac{c + BD - AD}{2(s-b)},$$

i.e.

$$\frac{(s-b)\rho_1}{r} = c + BD - AD. \tag{4}$$

Adding (3) and (4) we get

$$\left(\frac{r}{s-c} + \frac{s-b}{r} \right) \rho_1 = \frac{a+c-b}{2} s - b.$$

Hence

$$\rho_1 = \frac{r(s-b)(s-c)}{r^2 + (s-b)(s-c)}. \tag{5}$$

From (5) we get

$$r^2 = \frac{(s-c)(s-b)(s-c)}{s}$$

Hence

$$\rho_1 = \frac{rs}{2s-a} = \frac{F}{b+c},$$

where F is the area of $\triangle ABC$.

Analogously we get

$$\rho_2 = \frac{F}{c+a} \quad \text{and} \quad \rho_3 = \frac{F}{a+b}.$$

Then the inequality to be proved is equivalent to

$$F\left(\frac{1}{b+c} + \frac{1}{c+a} + \frac{1}{a+b}\right) \geq \frac{9}{4}r$$

i.e.

$$2(a+b+c)\left(\frac{1}{b+c} + \frac{1}{c+a} + \frac{1}{a+b}\right) \geq 9.$$

But this is equivalent to the inequality of Exercise 48. So

$$[(b+c)+(c+a)+(a+b)]\left(\frac{1}{b+c} + \frac{1}{c+a} + \frac{1}{a+b}\right) \geq 3^2$$

and the proof is complete.

83. Using the C-S-B inequality with $(a_1 + a_2)$ grouped together we see that

$$[(a_1+a_2)+a_3+a_4+\cdots+a_n]^2$$
$$\leq (1^2+1^2+\cdots 1^2)[(a_1+a_2)^2 + a_3^2 + \cdots + a_n^2].$$

Hence

$$\left(\sum_{i=1}^{n} a_i\right)^2 \leq (n-1)\left[\left(\sum_{i=1}^{n} a_i^2\right) + 2a_1 a_2\right].$$

So

$$\frac{1}{n-1}\left(\sum_{i=1}^{n} a_i\right)^2 \leq \left(\sum_{i=1}^{n} a_i^2\right) + 2a_1 a_2.$$

Using the inequality of the question

$$A < -\left(\sum_{i=1}^{n} a_i^2\right) + \frac{1}{n-1}\left(\sum_{i=1}^{n} a_i\right)^2$$
$$\leq -\left(\sum_{i=1}^{n} a_i^2\right) + \left(\sum_{i=1}^{n} a_i^2\right) + 2a_1 a_2 = 2a_1 a_2.$$

Similarly, $A \leq 2a_i a_j$ for $1 \leq i < j \leq n$.

84. Let $x = b+c-a$, $y = c+a-b$ and $z = a+b-c$. Now use $A(2) \geq G(2)$.

So

$$\frac{x+y}{2} \geq \sqrt{xy}, \quad \frac{y+z}{2} \geq \sqrt{yz}, \quad \frac{z+x}{2} \geq \sqrt{zx}.$$

Hence

$$\frac{1}{8}(x+y)(y+z)(z+x) \geq xyz.$$

So $abc \geq (b+c-a)(c+a-b)(a+b-c)$.

Multiplying out the right-hand side leads to the required result.

85. Let $a_1 = x_1 y_1 - z_1^2$, $a_2 = x_2 y_2 - z_2^2$ and $b = (x_1 + y_1)(x_2 + y_2) - (z_1 + z_2)^2$. The conditions of the problem give $a_1 > 0$, $a_2 > 0$. Since $x_1 > 0$, $x_2 > 0$ the fact that $a_1, a_2 > 0$ leads to $y_1, y_2 > 0$.

The inequality of the problem is equivalent to $\frac{1}{a_1} + \frac{1}{a_2} \geq \frac{8}{b}$. Since $b > 0$, this inequality becomes $(a_1 + a_2)b \geq 8a_1 a_2$.

Using the original equation for b we eventually find that

$$b = a_1 + a_2 + \frac{a_1 y_2}{y_1} + \frac{a_2 y_1}{y_2} + \left(\frac{z_1}{y_1} - \frac{z_2}{y_2} \right)^2 y_1 y_2 > 0.$$

Substituting for b from above we get the original inequality is equivalent to

$$(a_1 - a_2)^2 + (a_1 + a_2) \left(\frac{a_1 y_2}{y_1} + \frac{a_2 y_1}{y_2} \right)$$

$$+ (a_1 + a_2) \left(\frac{z_1}{y_1} - \frac{z_2}{y_2} \right)^2 \geq 4a_1 a_2 \qquad (*)$$

By the $A(2) \geq G(2)$ inequality,

$$a_1 + a_2 \geq 2\sqrt{a_1 a_2}, \text{ and } \frac{a_1 y_2}{y_1} + \frac{a_2 y_1}{y_2} \geq 2\sqrt{\frac{a_1 y_2}{y_1} \frac{a_2 y_1}{y_2}} = 2\sqrt{a_1 a_2}.$$

Hence the inequality $(*)$ holds since $(a_1 - a_2)^2 \geq 0$,

$$(a_1 + a_1) \left(\frac{z_1}{y_1} - \frac{z_2}{y_2} \right)^2 \geq 0, \text{ and } (a_1 + a_2) \left(\frac{a_1 y_2}{y_1} + \frac{a_2 y_1}{y_2} \right) \geq 4a_1 a_2.$$

This means therefore that the original inequality holds.

From the above argument, equality holds in $(*)$ if

$$a_1 = a_2, \quad \frac{a_1 y_2}{y_1} = \frac{a_2 y_1}{y_2} \text{ and } \frac{z_1}{y_1} = \frac{z_2}{y_2}.$$

Since the y_i are positive this means that $y_1 = y_2$. Hence $z_1 = z_2$. But since $a_1 = a_2$, $y_1 = y_2$ and $z_1 = z_2$, we have $x_1 = x_2$. Hence equality holds if and only if $x_1 = x_2$, $y_1 = y_2$ and $z_1 = z_2$.

86. Since $|f(x)| \leq 1$, there exists a smallest M such that $|f(x)| \leq M$. Since $f(x) \neq 0$ for some x, then $M > 0$.

Now suppose that there exists a such that $g(a) > 1$. Now $2|f(x)||g(a)| = |f(x + a) + f(x - a)| \leq |f(x + a)| + |f(x - a)|$ by the triangle inequality.

Hence $2|f(x)||g(a)| \leq 2M$.

So $|f(x)| \leq \frac{M}{|g(a)|} < M$.

But this contradicts the fact that M is the least upper bound of $f(x)$. Hence $|g(y)| \leq 1$ for all y.

87. By the C-S-B inequality, with $b_k = \sqrt{kx_k}$, and $c_k = \sqrt{k^5 x_k}$, we see that

$$\left[\sum_{k=1}^{5}(k^3 x_k)\right]^2 = a^4 \le \left(\sum_{k=1}^{5}(kx_k)\right)\left(\sum_{k=1}^{5}(k^5 x_k)\right) = a^4.$$

As equality holds, $c_k = \lambda b_k$. Hence $k^5 x_k = \lambda^2 k x_k$ or $x_k(k^4 - \lambda^2) = 0$ for $k = 1, 2, 3, 4, 5$. If all of the x_k are zero, then $a = 0$. Otherwise there exists an $x_k \ne 0$ in which case $\lambda = k^2$. However, $x_j = 0$ for all $j \ne k$ (since $x_k(k^4 - \lambda^2) = 0$ holds for $k = 1, 2, 3, 4, 5$). If $x_1 \ne 0$, then $x_1 = a$, $x_1 = a^2$ and $x_1 = a^3$ from the original equations. Hence $a = 1$. Similarly if the other $x_i \ne 0$ we get $a = 4, 9, 16, 25$.

88. Let s be the semi-perimeter and let $x = s - a$, $y = s - b$ and $z = s - c$. Then the given inequality is equivalent to

$$xy^3 + yz^3 + zx^3 \ge xyz(x + y + z).$$

Now applying the C-S-B inequality we see that

$$(xy^3 + yz^3 + zx^3)(z + x + y) \ge (y\sqrt{xyz} + z\sqrt{xyz} + x\sqrt{xyz})^2$$
$$= xyz(x + y + z)^2.$$

Hence the required inequality holds (since $x + y + z > 0$) with inequality being equality if $xy^3 = \lambda z$, $yz^3 = \lambda x$ and $zx^3 = \lambda y$. This leads to $x = y = z$ and so the triangle is equilateral.

89. Since x, y, z are non-negative and $x + y + z = 1$, at least one of these three numbers, say z, is less than or equal to a half. Thus the given expression

$$G = yz + zx + xy - 2xyz = z(x + y) + xy(1 - 2z)$$

written as a sum of non-negative terms is non-negative. For the least value of G, one of x, y must be 1 and the other 0. Thus $G \ge 0$.

By the $A(2)$, $G(2)$ inequality $x + y \ge 2\sqrt{xy}$, so

$$(1 - z)^2 = (x + y)^2 \ge 4xy.$$

Therefore

$$G - \frac{7}{27} = z(x + y) + xy(1 - 2z) - \frac{7}{27}$$
$$\le z(1 - z) + \frac{1}{4}(1 - z)^2(1 - 2z) - \frac{7}{27}$$
$$= -\frac{(3z - 1)^2(6z + 1)}{108} \le 0.$$

Thus $0 \le G \le 7/27$. The maximum of G is obtained only for $x = y = z = 1$.

90. Denote the lengths of the sides opposite A, B, C by a, b, c, respectively, and those of segments PD, PE, PF by x, y and z (see figure).

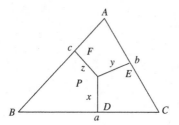

The area K of the triangle satisfies
$$2K = ax + by + cz. \tag{1}$$
We wish to minimize
$$\frac{a}{x} + \frac{b}{y} + \frac{c}{z} \tag{2}$$
subject to constraint (1). This can be done using the C-S-B inequality.
$$(u_1v_1 + u_2v_2 + u_3v_3)^2 \le (u_1^2 + u_2^2 + u_2^2)(v_1^2 + v_2^2 + v_3^2)$$
with $\sqrt{ax}, \sqrt{by}, \sqrt{cz}$ serving as u's and $\sqrt{\frac{a}{x}}, \sqrt{\frac{b}{y}}, \sqrt{\frac{c}{z}}$ as v's.

Thus
$$(a + b + c)^2 \le (ax + by + cz)\left(\frac{a}{x} + \frac{b}{y} + \frac{c}{z}\right) = 2K\left(\frac{a}{x} + \frac{b}{y} + \frac{c}{z}\right) \tag{3}$$
or
$$\frac{a}{x} + \frac{b}{y} + \frac{c}{z} \ge \frac{(a + b + c)^2}{2K}.$$

Equality holds if and only if the triples $(a/x, b/y, c/z)$ and (ax, by, cz) are proportional, i.e. if and only if $x = y = z$. Thus the minimum value of (2) occurs when P is the incentre of $\triangle ABC$.

Chapter 6

Combinatorics 3

6.1. Introduction

In this chapter I tackle the well known (to those who know it well) principle of inclusion–exclusion. It's a nifty way of doing some more counting. Using this principle as a theme I link in Stirling numbers of the second kind (there's a good name for a science fiction movie) Rook Polynomials and Catalan numbers. Both of these are interesting in their own right (or at least *I* think they are) and happen to have some use as well as we shall see.

These new ideas should have you running in circles at the sheer joy of the creative process or because of extreme frustration. If it's the latter remember no man is an island and too many cooks spoil the broth. If neither of those comments is useful, then vent your frustration on friends (maybe enemies might be better), parents and teachers. Together you will probably be able to sort out whatever tangled web I've woven for you. If not you can always look up the answer at the back of the chapter.

6.2. Inclusion–Exclusion

Very often it's better to count things indirectly rather than directly. For instance, if you wanted to know how many tyres are in the carpark, you'd do better to count the cars and multiply by 5, count the motorbikes and multiply by 2 and then add the two numbers you've got. This would be much quicker than walking round the carpark counting tyres.

In Chapter 2 of *First Step* and Chapter 1 of this book, we've already done some indirect counting. The aim of this section is to show you yet another technique called the ***Principle of Inclusion–Exclusion***. The aim of this method is to take carefully lavish overestimates and underestimates and refine the estimates in such a way that we end up with the correct answer.

So what we will do is to count everything in sight. Subtract too much away from this. Add in another overestimate. Subtract just a little more. Put in again... We'll continually include and exclude until we've done it just enough times to get the right answer.

This sounds more complicated than actually counting things directly, so let's look at an example. After the example you'll be convinced that it's more complicated than direct counting. However, I'll then give you some more difficult questions and then I think you'll see why inclusion–exclusion is a useful technique.

Example 1. How many positive integers less than 11 are not divisible by 2 or by 3?

The quick answer to this is to see that 1, 5, 7 are not divisible by 2 or 3. The answer is therefore just 3.

But now let's do this the inclusion–exclusion way. First of all there are 10 positive numbers less than 11, so let's first count 10 as the number we want.

In counting these 10 numbers I've clearly counted all the numbers divisible by 2 and all the numbers divisible by 3. So I should take 5 (number divisible by 2) and 3 (number divisible by 3) from the 10. This gives me 2. Finally, you should note that I have thrown away all the numbers that are divisible by both 2 *and* 3. There's only one of these, the number 6, so I must add 1 to the 2 I had above. This gives me 3, the answer we were looking for.

Exercises

1. By directly counting and also by including and excluding, find the number of positive integers less than 21 which are not divisible by 2 or by 3.
2. How many positive integers less than 31 are not divisible by 2 or by 5?
3. How many positive integers less than 101 are not divisible by 3 or by 5?
4. How many positive integers less than 1001 *are* divisible by 5 or by 7?

Actually a Venn diagram might help us to see what's going on a bit better. In Figure 6.1, A is the set of integers divisible by 2, B is the set of integers divisible by 3 and U is the set of positive integers less than 11.

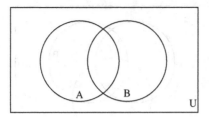

Figure 6.1.

Example 1 asks us to find the size of the set $U\backslash(A \cup B)$. Now

$$|U\backslash(A \cup B)| = |U| - |A \cup B|.$$

But $|A \cup B|$ is not just $|A| + |B|$ since $A \cap B \neq \emptyset$.

Clearly, $|A \cup B| = |A| + |B| - |A \cap B|$. This is because $A \cap B$ is counted once in $|A|$ and once in $|B|$. As it has already been counted twice, we subtract it off once to restore the balance.

So

$$|U\backslash(A \cup B)| = |U| - (|A| + |B| - |A \cap B|)$$
$$= |U| - (|A| + |B|) + (|A \cap B|).$$

This last statement checks with the lengthier argument of Example 1. We first count the whole set, U, we then subtract the bits A and B, and we finally add in what's in $A \cap B$.

The Venn diagram should help you to see why what we did in Example 1 actually works. You should also now see why we stopped at numbers divisible by 2 **and** 3. Oh, and questions like Exercise 4 should help you see that a direct count is not the most efficient way to do these problems.

Exercises

5. How many integers between 1 and 3,943 inclusive are not divisible by 11 or 13?
6. How many integers between 1 and 30 inclusive are not divisible by 2, 3 or 5? (Do this in two way).
7. Use a Venn diagram to justify the inclusion–exclusion method of Exercise 6.
8. How many integers between 1 and 9,264 are divisible by 4, 5 or 7?

The Venn diagram approach should help you to see how Exercise 6 works. It should also help you to see that there is nothing magic about the

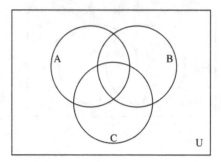

Figure 6.2.

numbers 30, 2, 3 or 5. The method will work whatever numbers you are using. So if you want to find all the integers from 1 to n which are not divisible by a, b or c, first draw the Venn diagram of Figure 6.2.

In this case let U be the integers from 1 to n, A be the subset divisible by a, B the subset divisible by b and C the subset divisible by c. So I want to find $|U \backslash (A \cup B \cup C)|$. The problem here, and in general, is that $|A \cup B \cup C| \neq |A| + |B| + |C|$, but it does start out like that. The trouble is that $|A| + |B| + |C|$ has counted $|A \cap B|$ once in $|A|$ and once in $|B|$. The same sort of thing goes for $|A \cap C|$ and $|B \cap C|$. So you might think that

$$|A \cup B \cup C| = |A| + |B| + |C| - (|A \cap B| + |B \cap C| + |C \cap A|).$$

However, there is a little problem. How many times have I counted $A \cap B \cap C$? Now $A \cap B \cap C$ was counted once in A, once in B and once in C and it's counted **out** once in $A \cap B$, once in $A \cap C$ and once in $B \cap C$. Three times in and three times out. So the set $A \cap B \cap C$ hasn't been counted at all! Let me put it back in.

$$|A \cup B \cup C| = |A| + |B| + |C| - (|A \cap B| + |B \cap C| + |C \cap A|)$$
$$+ |A \cap B \cap C|.$$

So I can now write down an expression for the thing I really wanted.

$$|U \backslash (A \cup B \cup C)| = |U| - (|A| + |B| + |C|) + (|A \cap B| + |B \cap C|$$
$$+ |C \cap A|) - |A \cap B \cap C|.$$

You should by now be getting a feel for how the including and excluding goes. You should also be reasonably convinced that it works. You should also know when to stop.

Exercises

9. How many integers from 1 to 365 inclusive are not divisible by 2, 3, 5 or 7?
10. How many integers between 397 and 4,328 inclusive are not divisible by 7, 9, 11 or 13?
11. Find an expression for $|A \cup B \cup C \cup D|$ in terms of the sets A, B, C, D, $A \cap B$, etc.
12. Find an expression for $|\bigcup_{i=1}^{n} A_i|$ in terms of the A_i and various intersections of the A_i.
13. Hence find an expression for $|U \setminus \bigcup_{i=1}^{n} A_i|$.

The Principle of Inclusion–Exclusion can be stated two ways, depending on whether or not you want a positive statement or a negative statement. Let's be positive first. How many objects in the set U have at least one of the properties $P_1, P_2, \ldots, P_n$? Let A_i be the subset of U with property P_i. The number of objects with the properties $P_1, P_2, \ldots, P_n$ is $|\bigcup_{i=1}^{n} A_i|$. From Exercise 12 we know that

$$\left| \bigcup_{i=1}^{n} A_i \right| = \sum_{i=1}^{n} |A_i| - \sum_{1 \leq i < j \leq n} |A_i \cap A_j| + \sum_{1 \leq i < j < k \leq n}$$
$$|A_i \cap A_j \cap A_k| \cdots + (-1)^{n+1} \left| \bigcap_{i=1}^{n} A_i \right|. \tag{1}$$

If, on the other hand, we want to know how many objects have none of these properties $P_1, P_2, \ldots, P_n$, then this is simply

$$\left| U \setminus \bigcup_{i=1}^{n} A_i \right| = |U| - \sum_{i=1}^{n} |A_i| + \sum_{1 \leq i < j \leq n} |A_i \cap A_j|$$
$$- \sum_{1 \leq i < j < k \leq n} |A_i \cap A_j \cap A_k| \cdots + (-1)^n \left| \bigcap_{i=1}^{n} A_i \right|. \tag{2}$$

To make the application of the Principle worthwhile, all those intersections on the right-hand side of the last two equations had better be easier to count than the union on the left-hand side.

Exercise

14. (a) In equation (1), how many terms are there with r intersecting sets, where $1 \leq r \leq n$?
    (b) In equation (1), what is the difference between the number of plus signs and the number of minus signs and why?

6.3. Derangements (Revisited)

In Chapter 1, when I looked at recurrence relations, I looked at the postman problem. You will recall that the new postman was so bad that he delivered every letter to the wrong house. With one letter per household, how many ways was this possible? The answer is the number of **derangements** of n things. We write this as p_n.

Let's tackle the problem afresh, especially as I have a new tool: Inclusion–Exclusion. How can I set up this problem for that method? Well, suppose there are four houses, 1, 2, 3, 4 and the postman has to deliver four letters, 1, 2, 3, 4 to the right houses but he does it all wrongly. No house gets the right letter (but all houses get one letter). We show below the nine ways this can be done. Hence $p_4 = 9$. See Table 1.

This can actually be done by counting in the following way.

Let A_1 = the number of deliveries with letter 1 delivered to house 1,

A_2 = the number of deliveries with letter 2 delivered to house 2,

A_3 = the number of deliveries with letter 3 delivered to house 3,

and A_4 = the number of deliveries with letter 4 delivered to house 4.

So what does $|\cup_{i=1}^{4} A_i|$ count here? Surely it's the number of ways the post can be delivered so that at least one letter is delivered to the correct house. That must mean that $p_n = |U \setminus \cup_{i=1}^{4} A_i|$ counts the required number of derangements. If I can calculate $|A_1|, |A_1 \cap A_2|, |A_1 \cap A_2 \cap A_3|$ and $|A_1 \cap A_2 \cap A_3 \cap A_4|$, then I should be in business.

For $|A_1|$, letter 1 goes to house 1, so there are three choices of letter left for house 2, then two choices for house 3 and finally one for house 4. Hence $|A_1| = 3! = 6$.

Similarly $|A_2| = |A_3| = |A_4| = 6$.

Table 1.

house 1	house 2	house 3	house 4
2	1	4	3
2	3	4	1
2	4	1	3
3	1	4	2
3	4	1	2
3	4	2	1
4	1	2	3
4	3	1	2
4	3	2	1

For $|A_1 \cap A_2|$, letter 1 goes to house 1 and letter 2 to house 2. This gives two choices for house 3 and one choice for house 4. Hence $|A_1 \cap A_2| = 2!$.

Similarly $|A_1 \cap A_3| = |A_1 \cap A_4| = |A_2 \cap A_3| = |A_2 \cap A_4| = |A_3 \cap A_4| = 2$.

For $|A_1 \cap A_2 \cap A_3|$, since three letters are already in place, the fourth one has to be in place too. So $|A_1 \cap A_2 \cap A_3| = 1$.

And similarly for $|A_1 \cap A_2 \cap A_4|, |A_1 \cap A_3 \cap A_4|$, and $|A_2 \cap A_3 \cap A_4|$.

Finally $|A_1 \cap A_2 \cap A_3 \cap A_4|$ is obviously 1.

So, putting all of that together gives me

$$\left| \bigcup\nolimits_{i=1}^{4} A_i \right| = 6 + 6 + 6 + 6 - (2 + 2 + 2 + 2 + 2 + 2)$$
$$+ (1 + 1 + 1 + 1) - 1 = 15.$$

Now $|U| = 24$ because this is all possible ways of delivering four letters to four houses with one letter per house. The number of derangements, p_4, is therefore

$$\left| U \setminus \bigcup\nolimits_{i=1}^{n} A_i \right| = 24 - 15 = 9.$$

Exercises

15. Find the number of ways of delivering three letters to three houses so that no house gets a correct letter. Do this in two ways.

16. Use the principle of inclusion–exclusion to find the number of derangements of five things.

What I've really been talking about here are special types of permutations. The nine sets of numbers in Table 1 above, are just permutations of the numbers 1, 2, 3, 4. These permutations have the extra property that no number is fixed or in its "natural" position. These derangements (permutations with no number fixed) compare with permutations like 1243 where both 1 and 2 are fixed. We know already that there are nine derangements of four numbers and 16 permutations which have at least one fixed number.

Exercises

17. Find the number of permutations of six numbers which fix just 1 and 2.

18. Find the number of permutations of n numbers which fix just 1 and 2.

19. Find the number of permutations of six numbers which fix exactly two numbers.

20. Find the number of permutations of n numbers which fix exactly two numbers.

21. Find the number of permutations of n numbers which fix exactly k numbers.

22. A permutation $(x_1, x_2, \ldots, x_{2n})$ of the set $\{1, 2, \ldots, 2n\}$, where n is a positive integer, is said to have property P if $|x_i - x_{i+1}| = n$ for at least one i in $\{1, 2, \ldots, 2n-1\}$.

Show that, for each n, there are more permutations with property P than without.

(IMO 1989, Question 6)

(Property P can be best described by looking at an example. Take $n = 3$, then $(1, 2, 4, 5, 3, 6)$ is the permutation which sends 1 to 1, 2 to 2, 3 to 4, 4 to 5, 5 to 3, and 6 to 6. Further $|1 - 2| = 1$ since 1 and 2 are one position apart. Similarly $|2 - 3| = 3, |3 - 4| = 2, |4 - 5| = 1$ and $|5 - 6| = 2$. Since $|2 - 3| = 3 = n$, then the permutation $(1, 2, 4, 5, 3, 6)$ has property P.

On the other hand, $(1, 3, 2, 4, 6, 5)$ does not have property P since $|x_i - x_{i+1}| \leq 2$ for all i).

6.4. Linear Diophantine Equations Again

If you can find Chapter 2 of *First Step* you will see how to find the number of solutions of $x + y + z = 15$, where x, y, z are non-negative integers.

Just in case you don't have access to this chapter, then recall that 0, 1 sequences with fifteen 1's and two 0's can be used quite effectively to solve this problem. The reason for this is that every such sequence (11101111101111111, for instance) represents a solution ($x = 3, y = 5, z = 7$ in this case). On the other hand, any solution ($x = 10, y = 0, z = 5$, for instance) can be represented as a binary sequence (11111111110011111 in this case).

Because there is a one-to-one map between solutions and sequences, counting the sequences gives the number of solutions. The number of solutions is therefore $^{17}C_2$.

Exercise

23. Find the number of solutions in non-negative integers of the following equations:

(i)　$x + y = 14$;　(ii) $x + y + z = 5$;　(iii) $w + x + y + z = 20$.

But what if we restrict ourselves to (strictly) positive integers? How many solutions in positive integers does $x + y = 4$ have? Clearly x can be 1, 2, or 3. So there are only three solutions.

Exercise

24. Use the principle of inclusion–exclusion to find the number of solutions in positive integers of the following equations.
    (i) $x + y = 14$; (ii) $x + y + z = 5$; (iii) $w + x + y + z = 20$.
    When possible, express your answer in terms of Binomial Coefficients.

There is actually a simpler method of solving problems like Exercise 24. For a start let's look at $x + y + z = 5$. Now

$$1 + 1 + 1 + 1 + 1 = 5. \tag{3}$$

Further, the required solutions are

$$x = 1, y = 1, z = 3; \quad x = 1, y = 2, z = 2; \quad x = 1, y = 3, z = 1;$$
$$x = 2, y = 1, z = 2; \quad x = 2, y = 2, z = 1; \quad x = 3, y = 1, z = 1.$$

We can get these solutions from (3) by noticing the equivalences

$$1 + 1 + 1 \oplus 1 \oplus 1; \quad 1 + 1 \oplus 1 + 1 \oplus 1; \quad 1 + 1 \oplus 1 \oplus 1 + 1;$$
$$1 \oplus 1 + 1 + 1 \oplus 1; \quad 1 \oplus 1 + 1 \oplus 1 + 1; \quad 1 \oplus 1 \oplus 1 + 1 + 1.$$

Every choice of two plus signs from equation (3) give us a solution to $x + y + z = 5$. Every solution of $x + y + z = 5$ gives a way of choosing two plus signs from equation (3). Hence there are as many solutions of $x + y + z = 5$ as there are ways of choosing two plus signs from four. Surely this can be done in ${}^4C_2$ ways, so the answer's 6.

Exercises

25. Find the number of positive integer solutions of the following equations:
    (i) $x + y + z = 20$; (ii) $w + x + y + z = 30$.
26. Find the number of positive integer solutions of the equation
    $$x_1 + x_2 + \cdots + x_n = n.$$
27. How many solutions in positive integers does the equation $x + y + z = 20$ have, if $x \geq 6$?
28. How many solutions, in integers greater than 5, does the equation $x + y + z = 20$ have?
29. How many solutions, in positive integers less than 10, does the equation $x + y + z = 20$ have?

30. Find the number of non-negative integral solutions of the equation $x + y = 21$, where $0 \le x \le 8$ and $0 \le y \le 9$.
31. Find the number of non-negative integral solutions of the equation $x + y + z = 25$, where $x \ge 4, y \le 15$ and $z \ge 2$.
32. Find the number of integral solutions of $x + y + z = 12$, where $-3 \le x \le 4, 2 \le y \le 11$, and $z \ge 3$.

6.5. Non-taking Rooks

From the heading, some of you may be expecting a story on black birds who are of a different ilk from the Jackdaw of Rheims (of poetry fame). On the other hand, 99.9% of you are already thinking of castles and chessboards. The latter group is more prosaic but also more correct.

The standard non-taking rook problem is to ask how many non-taking rooks can be placed on a chessboard. For those of you who have forgotten all the chess you ever knew, a rook (or castle) can only move along rows or columns of the chessboard. Since two rooks on a common row or column would be in a potential taking mode, then we can have at most eight non-taking rooks on the regular 8 by 8 board. Clearly by putting the eight castles all on one diagonal of the chessboard, none of them can get at any of the others. So eight rooks can be placed in mutually non-taking positions on a regular chessboard.

Notice how the following set of exercises leads to a generalisation of this idea.

Exercises

33. (a) In how many ways can three non-taking rooks be placed on a 3 by 3 chessboard?
    (b) Repeat (a) for four rooks on a 4 by 4 board.
    (c) Repeat (a) for eight rooks on the normal chessboard.
    (d) What about n rooks on an n by n board?
34. (a) In how many ways can two non-taking rooks be placed on a 3 by 3 board?
    (b) In how many ways can three non-taking rooks be placed on a 4 by 4 board?
    (c) Generalise (a) and (b).
35. Generalise Exercises 33 and 34. By this I mean, find the number of ways it is possible to put s rooks on an n by n board.

If you can't see how to do this immediately, try various values of *n* and *s* and try to find a pattern.

36. For those of you that don't have hang-ups about rectangular chessboards, find the number of ways of placing *s* non-taking rooks on an *m* by *n* board.

This all leads quite naturally on to rook polynomials. They are the generating functions (see Chapter 1) if you like, of the various rook numbers developed in the last exercises. After a little introduction to their basic ideas, I'll show you some useful problems that they can be used to solve.

We define the **rook polynomial** of the regular 8 by 8 chessboard C, to be

$$R(x, C) = r_0 + r_1 x + r_2 x^2 + r_3 x^3 + r_4 x^4 + r_5 x^5 + r_6 x^6 + r_7 x^7$$
$$+ r_8 x^8 = \sum_{i=0}^{8} r_i x^i,$$

where r_i is the number of ways of placing i non-taking rooks on the board C.

But if B is **any** type of chessboard whatsoever, we define its rook polynomial to be

$$R(x, B) = \sum_{i=0}^{n} r_i(B) x^i,$$

where $r_i(B)$ is the number of ways of placing i non-taking rooks on the board B. Further the number n is the maximum number of non-taking rooks that can be placed on B. To make life easy I'll always say that $r_0(B) = 1$. If you like there is only one way of placing no rooks on B. That way is to leave B empty.

Exercise

37. Find the rook polynomials for the board B, below.
    (i) a 2 by 2 board;   (ii) a 3 by 3 board;
   (iii) a 4 by 4 board;   (iv) a 1 by 2 board;
    (v) a 2 by 3 board;   (vi) a 3 by 4 board;
  (vii) a 4 by 5 board; (viii) a 4 by 6 board.

But if I can let my standards drop sufficiently for me to allow **rectangular** boards, why can't I have boards like the one in Figure 6.3?

What is $R(x, D)$, the rook polynomial of D? Is that really a problem? Swallow your pride you purists. All you have to do is to find $r_0(D), r_1(D),$ $r_2(D), r_3(D), r_4(D), \ldots$ Hmm. Surely we don't have to go past r_3. Since D has only three columns (or three rows) we can't put **four** non-taking rooks

Figure 6.3.

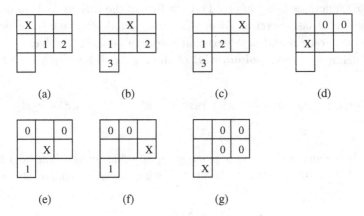

Figure 6.4.

on D. So what are the various numbers for this board? Let me do it systematically starting from no rooks and working up.

Now, as in any board, $r_0(D) = 1$.

There are seven squares on the board so there are seven ways of placing one non-taking rook on D. Hence $r_1(D) = 7$.

For two non-taking rooks we need to do a little work. Suppose I put the first rook in the position marked by a cross in Figure 6.4(a). Then the other rook can go in the squares marked 1 or 2.

A similar count can be done in Figure 6.4(b), where I put the first rook at position cross and the second rook in any of the positions marked 1, 2, or 3.

Figure 6.4(c) follows the same pattern. However, in Figure 6.4(d) I have zeros and no 1's or 2's. This is because, by placing the first rook at cross, the two obvious positions for the second rook (the two zeros) do not give *new* positions for the non-taking rooks. They simply repeat positions already counted in Figures 6.4(b) and 6.4(c).

In the same way we only get one new configuration in each of Figures 6.4(e) and 6.4(f). The potential positions marked with zeros were counted in Figures 6.4(a) and 6.4(c).

Altogether then, there are 10 possible ways of putting two rooks on this board. So $r_2(D) = 10$.

If I now work systematically I can find $r_3(D)$. This shouldn't be **too** hard.

Exercises

38. Determine $r_3(D)$, where D is the board of Figure 6.3.
    Hence determine $R(x, D)$, the rook polynomial of D.
39. Find rook polynomials for the boards below.

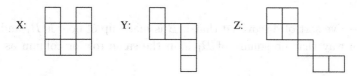

If you go back to the last rook polynomial in Exercise 39, you'll find that it factorizes. Indeed its factors are $1 + 4x + 2x^2, 1 + x$ and $1 + 2x$. There surely is some good reason for this. Have you seen any of these factors before?

You've certainly seen a board whose rook polynomial is $1 + 4x + 2x^2$. There's another board whose rook polynomial is $1 + 2x$.

If you can't remember seeing these polynomials have a sneak look back at Exercise 37. The 2×2 board gives you $1 + 4x + 2x^2$ and the 1×2 board gives $1 + 2x$. Is $1 + x$ a rook polynomial too? What is the rook polynomial of a 1×1 board?

Because Z is made up of three parts which don't have any influence on each other, it turns out that its rook polynomial is the **product** of the rook polynomials of its three parts. Does this generalise?

Exercises

40. Find rook polynomials for the boards below. Where possible factorize these polynomials into factors which are themselves rook polynomials. Is there a pattern?

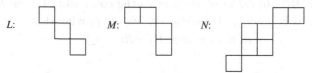

41. Find the rook polynomial of the board below and show that it is the product of the rook polynomial of a 2×2 board with the rook polynomial of a 3×3 board.

B: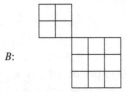

42. State a generalisation of the answers from the last three exercises. Why does this generalisation work?

So you've sort of discovered that if B is made up of boards B_1 and B_2 in such a way that no square of B_1 is in the same row or column as B_2, then

$$R(x, B) = R(x, B_1) \cdot R(x, B_2).$$

Boards for which no square of one is in the same horizontal row or vertical column as that of the other, are called ***non-interfering*** boards. Placing a rook in B_1 doesn't interfere with any rook placement in B_2.

Let's see how that works. Take the case of two non-interfering rooks where I want to find $r_2(B)$, say. Now I can put two rooks on B_1 in $r_2(B_1)$ ways. This gives $r_2(B_1) \cdot r_0(B_2)$ ways so far.

But then I can put one rook anywhere in B_1 and the other anywhere in B_2. These rooks will be non-taking because B_1 and B_2 have no common rows or columns. So altogether the number of ways of putting two rooks in B so that one is in B_1 and the other is in B_2 is $r_1(B_1) \cdot r_1(B_2)$.

Finally I can put no rooks in B_1 and two rooks in B_2. This contributes a total of $r_0(B_1) \cdot r_2(B_2)$ to the final $r_2(B)$ count.

So, altogether I'll have

$$r_2(B) = r_2(B_1) \cdot r_0(B_2) + r_1(B_1) \cdot r_1(B_2) + r_0(B_1) \cdot r_2(B_2).$$

But $r_2(B_1) \cdot r_0(B_2) + r_1(B_1) \cdot r_1(B_2) + r_0(B_1) \cdot r_2(B_2)$ is just the coefficient of x^2 in $R(x, B_1) \cdot R(x, B_2)$. Since this number is $r_2(B)$, then the coefficient of x^2 in $R(x, B_1) \cdot R(x, B_2)$ is the same as the coefficient of x^2 in $R(x, B)$.

Surely this is the first step along the way to proving that

$$R(x, B) = R(x, B_1) \cdot R(x, B_2)?$$

Exercises

43. (a) Show that the coefficient of x^3 is the same in $R(x, B)$ as it is in $R(x, B_1) \cdot R(x, B_2)$, where B_1 and B_2 are non-interfering boards.
    (b) Repeat (a) for the coefficient of x^4.
    (c) Repeat (a) for the coefficient of x^k, for $0 \le k \le n$, where n is the maximum number of non-taking rooks on the board B.

44. Prove $R(x, B) = R(x, B_1) \cdot R(x, B_2)$, where B_1 and B_2 together make up B and B_1 and B_2 have no rows or columns in common.

For ease of reference later, we will refer to the product rule we have produced as Property I.

Property I. *If B_1 and B_2 are non-interfering boards which together form the board B, then*

$$R(x, B) = R(x, B_1) \cdot R(x, B_2).$$

But determining rook polynomials can be a longwinded process if we are only armed with Property I and our fingers (to count on). Unfortunately there's not a lot more that we can do, although Property II does help a little.

Is there any way we can get $R(x, B)$ in some way from interfering boards contained in B? Let's try to find the rook polynomial of the board D in Figure 6.3.

I know that I've found $R(x, D)$ already but let me try to break D up a bit. In Figure 6.5, I show D, a marked square of D and two resulting smaller boards in.

The board D_1 is obtained from D by deleting the row and column which contains the marked square of D. The board D_2 is found by deleting just the marked square of D.

Now from Exercise 37 we know that $R(x, D_1) = 1 + 4x + 2x^2$ and $R(x, D_2) = 1 + 6x + 6x^2$. We also know from our earlier work that

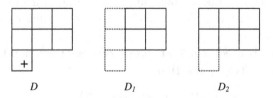

$$D \qquad\qquad D_1 \qquad\qquad D_2$$

Figure 6.5.

$R(x, D) = 1 + 7x + 10x^2 + 2x^3$. How can we get $R(x, D)$ by combining $R(x, D_1)$ and $R(x, D_2)$?

Exercises

45. Show that $R(x, D) = xR(x, D_1) + R(x, D_2)$. Give a plausible argument as to why this should be the case.

46. Consider the board X of Exercise 39.

    Take X and choose an arbitrary square t and form X_1 from X by deleting all the squares in the same row and column as t. Form X_2 from X by deleting the square t.

    Find $R(x, X_1)$, $R(x, X_2)$.

    Is $R(x, X) = xR(x, X_1) + R(x, X_2)$ or

    is $R(x, X) = xR(x, X_2) + R(x, X_1)$?

    Try to justify your answer.

47. Repeat the last exercise with the boards Y and Z of Exercise 39.

48. State, and, if possible, prove, a new property of rook polynomials.

So start with any board B. Choose some square t in B. Form B_1 from B by deleting all the squares in the same row and column as t. Form B_2 from B by deleting square t.

Property II. $R(x, B) = xR(x, B_1) + R(x, B_2)$.

If this indeed is a property, then it jolly well ought to have a proof. Maybe you've found the proof already in the last exercise. Suppose we want to find $r_3(B)$. When we put all three non-taking rooks on the board B, then either one of them occupies the square t or it doesn't.

In the former case, there must be two non-taking rooks on the part of B which has no square in the same row or column as B. But this part of B is B_1. We can put two non-taking rooks on B_1 in $r_2(B_1)$ ways.

In the latter case, none of the non-taking rooks is on the square t. There are therefore three non-taking rooks in the board B with t excluded. This sub-board of B is just B_2. We can place three non-taking rooks on B_2 in $r_3(B_2)$ ways.

Well either t is used or it isn't. So we can put three non-taking rooks on B in $r_2(B_1) + r_3(B_2)$ ways. Hence

$$r_3(B) = r_2(B_1) + r_3(B_2).$$

Exercises

49. (a) Show that $r_2(B) = r_1(B_1) + r_2(B_2)$.
    (b) Show that $r_4(B) = r_3(B_1) + r_4(B_2)$.
    (c) Find $r_k(B)$ in terms of appropriate numbers from B_1 and B_2, for $0 \le k \le n$, where n is the maximum number of non-taking rooks on B.

50. Prove Property II.

51. Use Properties I and II to check your answers to Exercise 37.

6.6. The Board of Forbidden Positions

Doesn't that title have an ominous ring to it? OK, so in how many ways can a postman post four letters in four different, wrong, letterboxes? Sorry to raise this old chestnut again. You ought to immediately respond with $p_4 = 4! - 4! + \frac{4!}{2!} - \frac{4!}{3!} + \frac{4!}{4!} = 9$ (see Section 6.3). But let me relate this to chessboards. Look at Figure 6.6.

I want to do this in a back to front way so bear with me. I'm going to find the number of derangements by first finding all the ways of filling the 4×4 board above so that none of the squares marked with a cross are used. Let me call the board formed by the crossed squares the **board of forbidden positions** — it's where I'm not allowed to go.

But let me go there anyway, because the rook polynomial of this crossed square board is easy. Obviously a single square has polynomial $1 + x$, so, by Property I, this board of forbidden positions **(bfp)** in Figure 6.6 has rook polynomial $(1 + x)^4$. Now

$$(1 + x)^4 = 1 + 4x + 6x^2 + 4x^3 + x^4.$$

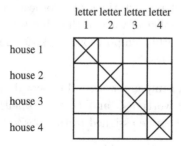

Figure 6.6.

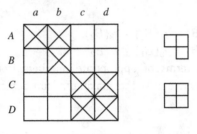

Figure 6.7.

So for the bfp, $r_0 = 1, r_1 = 4, r_2 = 6, r_3 = 4, r_4 = 1$.

Now look at this.

$$4! \cdot r_0 - 3! \cdot r_1 + 2! \cdot r_2 - 1! \cdot r_3 + 0! \cdot r_4 = 4! - 3! \cdot 4 + 2! \cdot 6 - 1! \cdot 4 + 0! \cdot 1$$

$$= 4! - 4! + \frac{4!}{2!} - \frac{4!}{3!} + \frac{4!}{4!}$$

$$= 9.$$

It would seem that somehow we can find p_4 by looking at the **bfp** in Figure 6.6 and doing some weird calculations of the form

$$\sum_{k=0}^{n} (-1)^k (n-k)! r_k.$$

Exercises

52. Find p_3, p_5 and p_6 by the formula of p_n **and** by the **bfp** method.

53. Find p_n by the **bfp** method.

The reason this strange formula works is simply because it is the inclusion–exclusion argument of Section 6.3 in disguise. The term $(n-k)! r_k$ is simply the number of permutations of n objects with k fixed points. In each case r_k is the number of ways of choosing k from n.

But if this board of forbidden positions works for derangements, maybe it'll work for other things too.

Example 2. In a given company, worker A cannot do jobs a and b, worker B cannot do job b, and workers C and D can do neither c nor d. In how many ways can the workers be assigned to the jobs?

First of all let me make the problem one that has something to do with **bfp**'s. To do this I've drawn up the 4×4 board in Figure 6.7 and placed crosses where jobs are not possible.

So, for instance, there is a cross in the (B, b) square since B can't do job b.

Let me now find the rook polynomial for the "crossed" board, the board of forbidden positions, if you like. By Property I, the rook polynomial we want is the product of the two small boards in the diagram above. So the rook polynomial is

$$(1 + 3x + x^2)(1 + 4x + 2x^2) = 1 + 7x + 15x^2 + 10x^3 + 2x^4.$$

So $r_0 = 1, r_1 = 7, r_2 = 15, r_3 = 10, r_4 = 2$. Hence

$$\sum_{k=0}^{n} (-1)^k (n - k)! r_k = 4! - 3! \cdot 7 + 2! \cdot 15 - 1! \cdot 10 + 0! \cdot 2 = 4.$$

Is this indeed, the required number? Let's work it out from scratch.

Clearly if B is given job a, then we can't assign jobs to C and D. So we have the assignments (i) $A : c, B : d, C : a, D : b$; (ii) $A : c, B : d, C : b, D : a$; (iii) $A : d, B : c, C : a, D : b$; (iv) $A : d, B : c, C : b, D : a$. Precisely four!

Exercises

54. Amy, Baker, Charleen and Dick have been elected as the social committee for Form 5G. Charleen and Dick refuse to be chairperson, Amy won't be treasurer and Baker says he's not going to be the secretary or the fund raiser. Charleen also doesn't want to be secretary. In how many ways can the four positions on the social committee be filled?

55. A computer dating service is trying to match Ann, Belinda, Carol, Deirdre and Elaine with Alan, Brian, Charlie, David and Ed. The service decides that Ann is unsuitable for Brian and Charlie, that Alan is unsuitable for Carol as are Brian and Charlie, that David and Elaine are unsuitable for each other and that the same is true for Deirdre and Ed and for Belinda and Brian. How many compatible dates are there?

56. Why does the formula $\sum_{k=0}^{n} (-1)^k (n - k)! r_k$ work?

Let $E = \sum_{k=0}^{n} (-1)^k (n - k)! r_k$. What we are trying to do in all these rook polynomial counting-type problems is to count special permutations. One of the answers, $A : c, B : d, C : a, D : b$, from Example 2 is equivalent to the permutations which interchange 1 and 3 as well as 2 and 4. To see this, change both A, B, C, D and a, b, c, d, to 1, 2, 3, 4. Then $A : c, C : a$ is equivalent to interchanging 1 and 3 and $B : d, D : b$ is equivalent to interchanging 2 and 4.

E is then trying to count all the permutations of n objects, subject to some restrictions — those imposed by the board of forbidden positions. This is clearly all possible permutations minus those which have some restrictions caused by the **bfp**.

The form of the inclusion–exclusion relation we're looking at is given in equation (2) on page 171. This is

$$\left| U \setminus \bigcup_{i=1}^{n} A_i \right| = |U| - \sum_{i=1}^{n} |A_i| + \sum_{1 \leq i < j \leq n} |A_i \cap A_j|$$
$$- \sum_{1 \leq i < j < k \leq n} |A_i \cap A_j \cap A_k| \cdots + (-1)^n \left| \bigcap_{i=1}^{n} A_i \right|.$$

In this application, A_i is the number of permutations with i in a forbidden position. So $|A_i| = (n-1)! \times \{$number of forbidden squares in the i-th column of the **bfp**$\}$.

Further $\sum_{i=1}^{n} |A_i| = (n-1)! \times \{$number of squares in **bfp**$\} = (n-1) \cdot r_1$.

How about $|A_i \cap A_j|$? This is $(n-2)! \times \{$number of ways of choosing squares in the i-th and j-th column of the **bfp** each in a different row$\}$.

So

$\sum_{1 \leq i < j < n} |A_i \cap A_j| = (n-2)! \times \{$number of ways of choosing squares in any 2 columns of the **bfp**, each in a different row$\} = (n-2)! \cdot r_2$.

Similarly $\sum_{1 \leq i < j < k \leq n} |A_i \cap A_j \cap A_k| = (n-3)! \times \{$number of ways of choosing squares in any 3 columns of the **bfp**, each in a different row$\} = (n-3)! \cdot r_3$.

But $|U| = n!$ because this counts all possible permutations. And $n! = n! \cdot r_0$, since $r_0 = 1$ by convention for all boards. Hence

$$E = \left| U \setminus \bigcup_{i=1}^{n} A_i \right| = |U| - (n-1)! r_1 + (n-2)! r_2 + \cdots + (-1)^n r_n.$$
$$E = \left| U \setminus \bigcup_{i=1}^{n} A_i \right| = \sum_{k=0}^{n} (-1)^k (n-k)! r_k.$$

Just what all the fuss was about?

Exercises

58. How many ways can jobs be assigned if A and F can do jobs $1, 2, 5, 6$, B and E can do jobs $1, 3, 4, 6$ and C and D can do jobs $2, 3, 4, 5$?

59. Let $R_n^m(x)$ be the rook polynomial of an $m \times n$ rectangular board. Show that
    (a) $R_n^m(x) = R_{n-1}^m(x) + mx R_{n-1}^{m-1}(x)$;
    (b) $\frac{d(R_n^m(x))}{dx} = nm R_{n-1}^{m-1}(x)$.
    Hence find $R_5^2(x)$.

60. Find two different boards which have the same rook polynomial.

6.7. Stirling Numbers

James Stirling clearly had no close encounters of the third kind because he only bequeathed us numbers of the first and second kind. (Although maybe he did have an encounter of the third kind before he could invent Stirling numbers of the third kind.)

Anyway, the number of ways of partitioning a set of size n into precisely r non-empty subsets is said to be the **Stirling number of the second kind** and is denoted by $S(n, r)$.

Clearly $S(n, 1) = 1$. There is only one way of breaking a set of size n into 1 subset.

But $S(4, 2)$ is not so clear. So I'll work it out. Suppose I have the set $\{a, b, c, d\}$ and I want to partition it into 2 subsets. Then this can be done as follows:

$$a, bcd \quad b, acd \quad c, abd \quad d, abc \quad ab, cd \quad ac, bd \quad ad, bc.$$

So $S(4, 2) = 7$.

But what is the use of Stirling numbers of the second kind and what are Stirling numbers of the first kind?

Exercises

61. Find the following Stirling numbers of the second kind.
    (i) $S(4, 1)$; (ii) $S(4, 3)$; (iii) $S(4, 4)$; (iv) $S(4, 0)$.
62. Determine
    (i) $S(n, n)$; (ii) $S(n, n - 1)$.
63. (a) Find a and b in the following equation
    $$S(4, 2) = a \cdot S(3, 2) + b \cdot S(3, 1).$$

    Now do it other than by first calculating all the Stirling numbers involved.
    (b) Repeat (a) with $S(5, 2) = a \cdot S(4, 2) + b \cdot S(4, 1)$.
    (c) Generalise (a) and (b).

In Chapter 1, I introduced the idea of a recurrence relation. Exercise 63 should suggest that $S(n, r)$ satisfies a recurrence relation of this form

$$S(n, r) = a \cdot S(n - 1, r) + b \cdot S(n - 1, r - 1).$$

But what is a and what is b?

Before I answer that question, let's go back to $S(5, 2)$. In Exercise 63(b) you found that

$$S(5, 2) = 2S(4, 2) + S(4, 1).$$

To do that without determining $S(5, 2), S(4, 2)$ and $S(4, 1)$, suppose that the elements of the 5-set are a, b, c, d, e. Now $\{a\}$ is either one of the two sets or it is in a set with at least two elements.

There is only one way $\{a\}$ can be one of the two sets. But this one way can be thought of as the number of ways of partitioning the elements b, c, d, e into **one** set. In other words $S(4, 1)$.

On the other hand, if a is in a larger set, then I'll forget about it for a minute and concentrate on the four elements b, c, d, e left behind. These can be divided into **two** sets in $S(4, 2)$ ways.

However, for each such partitioning of b, c, d, e into 2 sets we get **two** partitions of a, b, c, d, e into 2 sets. This is because a can be added to either of the sets.

So $S(5, 2) = 2S(4, 2) + S(4, 1)$.

Exercises

64. (a) Show that $S(5, 3) = 3S(4, 3) + S(4, 2)$.
    (b) Show that $S(n, r) = rS(n - 1, 1) + S(n - 1, r - 1)$.
65. Use the last exercise to determine $S(6, r)$ for all possible values of r.
66. Show that there are $2! \cdot S(3, 2)$ ways of putting three differently coloured balls into two different shaped boxes so that no box is left empty.
67. (a) In how many ways can four differently coloured balls be put into exactly three differently shaped boxes?
    (b) If $m \geq n$, how many ways can m labelled objects be put into exactly n labelled containers?

The reason that I am talking about Stirling numbers of the second kind is that an expression can be found for them using, naturally, inclusion–exclusion.

Now in Exercise 67 you discovered that there are $n! \cdot S(m, n)$ ways of putting m labelled objects into n labelled containers. To use inclusion–exclusion I note that it is easy to find $B(r)$ the number of ways of putting m labelled objects into n containers if we leave at least r containers empty. This is because we can choose the r empty containers in ^nC_r ways. Then the remaining $n - r$ containers can be filled in $(n - r)^m$ ways. This is because the first object can be put into $n - r$ containers, and the second object can be put into $n - r$ containers and so on for all m objects. Multiplying each $n - r$ together m times gives the required count.

But the total number of ways of putting m labelled objects into n labelled containers is

$$B(0) - B(1) + B(2) - B(3) + \cdots \pm B(n)$$

by inclusion–exclusion.

So $\quad n! \cdot S(m, n) = {}^nC_0(n)^m - {}^nC_1(n - 1)^m + {}^nC_2(n - 2)^m - \cdots$

$$= \sum_{r=0}^{n} (-1)^r \ {}^nC_r(n - r)^m.$$

Hence

$$S(m, n) = \sum_{r=0}^{n} (-1)^r \frac{(n - r)^m}{r!(n - r)!}. \tag{4}$$

Exercises

68. Find $S(4, r)$ for $r = 0, 1, 2, 3, 4$ using equation (4).
69. Use equation (4) to establish the recurrence relation
$$S(m, n) = n \cdot S(m - 1, n) + S(m - 1, n - 1).$$
70. The Stirling numbers of the first kind $S'(m, n)$ are the coefficients of x^n in the expansion of $x(x - 1) \cdots (x - m + 1)$. In other words
$$x(x - 1) \cdots (x - m + 1) = \sum_{n=0}^{m} (-1)^n S'(m, n) x^n.$$
   (a) Find $S'(4, 1)$, $S'(5, 2)$ and $S'(6, 2)$.
   (b) Show that $S'(m, m - 1) = \frac{1}{2}m(m - 1)$.
   (c) Show that $S'(m, n) = (m - 1) \cdot S'(m - 1, n) + S'(m - 1, n - 1)$.

6.8. Some Other Numbers

At first sight, the point of this section appears to be to introduce a new set of numbers and explore its properties. However, there is a further underlying combinatorial idea. That is that to count a set of objects we may approach it head on and find its size directly or we may find another set whose size we know and show that the two sets are equinumerous.

I came across the following interesting problem many years ago when it appeared in the nrich web site run by Cambridge University for bright maths students (http://nrich.maths.org/714). They called it One Basket or Group Photo.

The Photo Problem. *A school photographer is taking a photograph of the two basketball teams. She has to arrange ten people, all of different heights, in two rows of five, one behind the other. Each person at the back must be taller than the person directly in front of them. Along the rows the heights must increase from left to right.*

In how many ways can two, four or six people to be arranged in this way for a photo, or eight people? In how many ways can the ten team members be arranged like this for the photo to be taken?

You may even like to generalise the problem to twelve people or to any specified even number.

It's probably too hard to try to tackle 10 basketballers straight away. There seem to be too many possibilities to keep track of them. So it might be worth using a common problem solving strategy and that is to try a simpler/smaller case. Of course, there's maybe no point in starting too simply. After all it should be clear that with two basketballers, there's only one photo. Does it get more interesting with four basketballers though?

Exercises

71. See what you can do with four, six and eight basketballers.
    Does that give you sufficient ideas, along with being systematic, to get the 10 basketballer case?

72. Do those numbers look like anything that you have seen before? Can you generalise?

73. The **Change Problem** can be found in the 1973 book "Mathematical Gems" by Ross Honsberger. It is about a line of $2n$ people waiting for admission to a theatre.
    Change Problem. *The price of admission is 50 cents and n of the people have the exact change while the other n have one dollar bills; thus each person either provides a unit of change for the cashier's later use $(+1)$ or uses up a unit of her change (-1). The question is "In how many ways can the patrons be lined up so that a cashier who begins with no change of her own, is never stuck for change?"*

74. **The Triangulation Problem.** A regular n-gon, can be divided up into $n-2$ triangles by joining vertices of the polygon. A given vertex may be joined to more than one other vertex. The result of joining up the vertices is called a ***triangulation***. How many different ways are there of triangulating a regular n-gon in this way?

You should note that the following arrangements for $n = 6$ are considered to be different, even though rotating one hexagon may give you the same triangular pattern as some other.

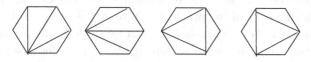

Having worked on those problems for a while you should start to see some interesting relations. So much so that you might wonder if there are any other situations that are counted by the same numbers.

Let's look at the **Bracket Problem**. How many ways can you put n pairs of brackets (a pair of a left and a right bracket) in an expression so that pairs of brackets match up? Note that we can't legally have the situation (()))(, so the number of left brackets appearing at any stage never exceeds the number of right brackets. (Does that remind you of anything?)

To see what I mean let me look at a few small cases.

$$n = 1: \quad ()$$
$$n = 2: \quad (()) \qquad\qquad ()()$$
$$n = 3: \quad ((())) \quad (()()) \quad (())() \quad ()(()) \quad ()()()$$

It's almost certain that I'm going to get 14 legitimate ways to arrange four pairs of brackets. I seem to be heading for the same sequence of numbers again.

Exercises

75. Show that there are indeed 14 ways of correctly placing four pairs of brackets.
76. If the problems that I have mentioned so far do indeed all count the same numbers, then there must be a one-to-one correspondence between them. Can you find any of these relations?

The numbers that we have been looking at here are called the *Catalan numbers*. I will represent them by L_n. (The usual practice is to use C_n but this notation can easily get confused with the Binomial Coefficients so I've chosen the unusual L instead of C.) From our experience so far, then, we have

$$L_1 = 1, \quad L_2 = 2, \quad L_3 = 5, \quad L_4 = 14 \quad \text{and} \quad L_5 = 42.$$

For a more complete list of Catalan numbers, see the On-Line Encyclopedia of Integer Sequences (http://oeis.org/A000108).

What I want to do in the rest of this section is to provide a formula for L_n, in general, look at some relationships between Catalan numbers and establish that each of the problems I have introduced, is in fact solved by the Catalan numbers. There might even be space to allow some more problems whose solutions are Catalan numbers. But, for now, back to Catalan numbers themselves.

I'll define the n-th Catalan number as:

$$L_n = \frac{(2n)!}{(n+1)!n!} \quad \text{for } n \geq 0.$$

But it can be written in other ways too.

$$L_n = \frac{1}{n+1} {}^{2n}C_n = {}^{2n}C_n - {}^{2n}C_{n+1} = \frac{1}{n+1} \sum_{i=0}^{n} ({}^{n}C_i)^2 \quad \text{for } n \geq 0. \quad (1)$$

The Catalan numbers also satisfy the recurrence relationships:

$$L_{n+1} = \frac{2(2n+1)}{n+2} L_n \quad \text{with } L_0 = 1 \quad\quad (2)$$

and

$$L_{n+1} = \sum_{i=0}^{n} L_i L_{n-i} \quad \text{with } L_0 = 1. \quad\quad (3)$$

Exercises

77. Show that the various equalities in equation (1) above, hold.
78. Prove that equation (2) holds for L_n.
79. Prove that the Catalan numbers count the Triangulation Problem of Exercise 74.

I'm now going to outline a proof that the Change Problem and the Bracket Problem involve equi-numerous objects. Let me think about the Change Problem first. Suppose that I think of 50c being given to the cashier as being a $+1$, and 50c being given back to a customer as being -1. Then I want to find all the ways of giving change so that the sum of the $+1$s and -1s up to any point is not negative. The key feature of these sequences is that at any point, there are never more -1s than there are $+1$s. But a similar thing is true for the brackets. We have strings of left and right brackets but at no point do we have more right brackets than left brackets. Hence a mapping from "$+1$" to "(" and "-1" to ")", is one-to-one.

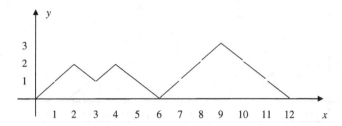

Figure 6.8. A graph of change.

Exercises

80. Prove the equivalence of the Bracket Problem and the Photo Problem.
81. Find other applications of the Catalan numbers.

The key thing now is to prove that one of these problems is counted by the Catalan numbers. I'll choose to look at the Change Problem because it involves an interesting argument. First though, the general set up of the proof. I'm going to show that there are $^{2n}C_n - {}^{2n}C_{n+1}$ ways to get the change. But I'm going to show the various possibilities of change graphically. From the graphs I'll over-count to get $^{2n}C_n$. The trick then is to see where the extra $^{2n}C_{n+1}$ graphs came from.

Now to construct the graphs. Let me think of 50c being given to the cashier as being a $+1$, and 50c being given back to a customer as being -1. Then I want to find all the ways of giving change so that the sum of the $+1$s and -1s up to any point is not negative. We can represent this on a graph by starting at $(0, 0)$ and moving up a unit and along a unit for $+1$ and down a unit and along a unit for -1. All such graphs should end up at the point $(2n, 0)$. I've shown one of these in Figure 6.8, with $n = 6$.

It turns out to be easy to count all of these graphs. There are simply $^{2n}C_n$ of them. This is just the number of ways of choosing n "$+1$s", say from a collection of $2n$ objects.

I would hope that none of these graphs ever go below the x-axis but I can't guarantee this. So the excess in my count must just be the number of graphs that go below the x-axis at some point. I show one of these in Figure 6.9 (the continuous line).

Now I notice an interesting fact. All of the graphs that go below the x-axis have to pass through the line $y = -1$. So I'm going to follow the course of a graph that is not one that reflects the change situation that we are counting. It has to go through the line $y = -1$ for a first time. Before that it is totally above or on the x-axis. Suppose that the graph hits

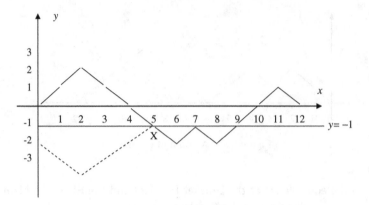

Figure 6.9.   A non-change graph and its reflection.

the line $y = -1$ at the point X. Then reflect all of the graph to the **left** of X about the line $y = -1$ as I have done in Figure 6.9 (the broken line). Since the original graph started at $(0, 0)$, the reflected one has to start at $(0, -2)$.

I now want to note that there is a one-to-one correspondence between all the graphs that are not change graphs and all the graphs from $(0, -2)$ to $(2n, 0)$. Clearly two non-change graphs that differ on the first section that ends at the x-axis lead to two different reflected graphs. What's more two non-change graphs that have the same first section to the x-axis, finish differently and so lead to different reflected graphs. And, similarly, two different reflection graphs lead to two different non-change, unreflected graphs.

Finally, the reflected graphs consist of $n + 1$ "+1s" and $n - 1$ "−1s". There have to be $^{2n}C_{n+1}$ of these and our count is complete.

Because we know that the Bracket, Change and Photo Problems are equivalent, we now know that the Bracket and Photo Problems generate the Catalan numbers in the same way that the Change Problem does.

6.9. Solutions

1. $7 = 20 - (10 + 6) + 2.$
2. $12 = 30 - (15 + 6) + 3.$
3. $53 = 100 - (33 + 20) + 6.$
4. $314 = 10000 - [10000 - (200 + 142) + 28].$
5. $3308 = 3943 - (358 + 303) + 26.$
6. $8 = 30 - (15 + 10 + 6) + (5 + 2 + 3) - 1.$

7.

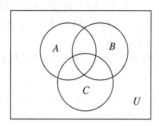

Let $U = \{1, 2, \ldots, 30\}$, A be the numbers divisible by 2, B be the numbers divisible by 3 and C be the numbers divisible by 5. Then

$$|U \backslash (A \cup B \cup C)| = |U| - (|A| + |B| + |C|)$$
$$+ (|A \cap B| + |B \cap C| + |C \cap A|) - |A \cap B \cap C|.$$

8. $(2316 + 1852 + 1323) - (463 + 264 + 330) + 66 = 4500$.

9. $365 - (192 + 121 + 73 + 52) + (60 + 36 + 26 + 24 + 17 + 10)$
$- (12 + 8 + 5 + 3) + 1 = 73$.

10. $[4328 - (618 + 480 + 393 + 332) + (68 + 56 + 47 + 43 + 36 + 30) -$
$(6 + 5 + 3) + 0] - [396 - (56 + 44 + 36 + 30) + (11 + 5 + 4 +$
$4 + 3 + 2) - (0 + 0 + 0) + 0] = 2512$.

11. $|A \cup B \cup C \cup D| = |A| + |B| + |C| + |D|$
$-(|A \cap B| + |A \cap C| + |A \cap D| + |B \cap C| + |B \cap D| + |C \cap D|)$
$+(|A \cap B \cap C| + |A \cap B \cap D| + |A \cap C \cap D| + |B \cap C \cap D|)$
$-|A \cap B \cap C \cap D|.$

12. $|\cup_{i=1}^{n} A_i| = \sum_{i=1}^{n} |A_i| - \sum_{1 \le i < j \le n} |A_i \cap A_j|$
$+ \sum_{1 \le i < j < k \le n} |A_i \cap A_j \cap A_k| - \cdots + (-1)^{n+1} |\cap_{i=1}^{n} A_i|.$

So $|U - \cup_{i=1}^{n} A_i| = |U| - |\cup_{i=1}^{n} A_i|$. Complete the rest using the above expression for $|\cup_{i=1}^{n} A_i|$.

13. (a) $^{n}C_r$ — the number of ways of choosing r things from n things.
(b) The number is $\sum_{r=1}^{n} (-1)^{r+1} \, {}^{n}C_r = 1 - (1-1)^n = 1$.

(See Chapter 2 of *First Step* for the expansion of $(1-x)^n$ via the Binomial Theorem.)

14. $p_3 = |U| - (|A_1| + |A_2| + |A_3|) + (|A_1 \cap A_2| + A_1 \cap A_3| + |A_2 \cap A_3|) - |A_1 \cap A_2 \cap A_3| = 6 - (2 + 2 + 2) + (1 + 1 + 1) - 1 = 2$.

15. $p_5 = |U| - [(|A_1| + |A_2| + |A_3| + |A_4| + |A_5|) - (|A_1 \cap A_2| + |A_1 \cap A_3|$
$+ |A_1 \cap A_4| + |A_1 \cap A_5| + |A_2 \cap A_3| + |A_2 \cap A_4| + |A_2 \cap A_5| + |A_3 \cap A_4|$
$+ |A_3 \cap A_5| + |A_4 \cap A_5|) + (|A_1 \cap A_2 \cap A_3| + |A_1 \cap A_2 \cap A_4|$
$+ |A_1 \cap A_2 \cap A_5| + |A_1 \cap A_3 \cap A_4| + |A_1 \cap A_3 \cap A_5|$

$$+|A_1 \cap A_4 \cap A_5| + |A_2 \cap A_3 \cap A_4| + |A_2 \cap A_3 \cap A_5| + |A_2 \cap A_4 \cap A_5|$$
$$+|A_3 \cap A_4 \cap A_5|) - (|A_1 \cap A_2 \cap A_3 \cap A_4| + |A_1 \cap A_2 \cap A_3 \cap A_5|$$
$$+|A_1 \cap A_2 \cap A_4 \cap A_5| + |A_1 \cap A_3 \cap A_4 \cap A_5| + |A_2 \cap A_3 \cap A_4 \cap A_5|)$$
$$+|A_1 \cap A_2 \cap A_3 \cap A_4 \cap A_5|]$$
$$= 5! - (^5C_1 4! - {}^5C_2 3! + {}^5C_3 2! - {}^5C_4 1! + {}^5C_5 0!) = 44.$$

16.

$$p_n = \left| U \setminus \bigcup_{i=1}^{n} A_i \right| = |U| - ({}^nC_1(n-1)! - {}^nC_2(n-2)! + \cdots)$$

$$= |U| + \sum_{r=1}^{n} (-1)^r \, {}^nC_r(n-r)!$$

$$= n! - n! + \frac{n!}{2!} - \frac{n!}{3!} + \frac{n!}{4!} - \cdots + (-1)^n \frac{n!}{n!}$$

$$= n! \left\{ 1 - 1 + \frac{1}{2!} - \frac{1}{3!} + \frac{1}{4!} - \cdots + (-1)^n \frac{1}{n!} \right\}.$$

This is precisely the expression for p_n obtained in the last exercise in Chapter 1.

17. We have all possible permutations of the four objects 3, 4, 5, 6. There are 24 such permutations.

18. $(n-2)!$.

19. Let A_i be the number of permutations which fix i. The permutations fixing exactly two numbers are
$$(|A_1 \cap A_2| + |A_1 \cap A_3| + |A_1 \cap A_4| + |A_1 \cap A_5| + |A_1 \cap A_6| + |A_2 \cap A_3|$$
$$+|A_2 \cap A_4| + |A_2 \cap A_5| + |A_2 \cap A_6| + |A_3 \cap A_4| + |A_3 \cap A_5| + |A_3 \cap A_6|$$
$$+|A_4 \cap A_5| + |A_4 \cap A_6| + |A_5 \cap A_6|) - 3(|A_1 \cap A_2 \cap A_3| + |A_1 \cap A_2 \cap A_4|$$
$$+|A_1 \cap A_2 \cap A_5| + |A_1 \cap A_2 \cap A_6| + |A_2 \cap A_3 \cap A_4| + |A_2 \cap A_3 \cap A_5|$$
$$+|A_2 \cap A_3 \cap A_6| + |A_2 \cap A_4 \cap A_5| + |A_2 \cap A_4 \cap A_6| + |A_2 \cap A_5 \cap A_6|$$
$$+|A_3 \cap A_4 \cap A_5| + |A_3 \cap A_4 \cap A_6| + |A_3 \cap A_5 \cap A_6| + |A_4 \cap A_5 \cap A_6|)$$
$$+6(|A_1 \cap A_2 \cap A_3 \cap A_4| + |A_1 \cap A_2 \cap A_3 \cap A_5| + |A_1 \cap A_2 \cap A_3 \cap A_6|$$
$$+|A_1 \cap A_2 \cap A_4 \cap A_5| + |A_1 \cap A_2 \cap A_4 \cap A_6| + |A_1 \cap A_2 \cap A_5 \cap A_6|$$
$$+|A_1 \cap A_3 \cap A_4 \cap A_5| + |A_1 \cap A_3 \cap A_4 \cap A_6| + |A_1 \cap A_3 \cap A_5 \cap A_6|$$
$$+|A_1 \cap A_4 \cap A_5 \cap A_6| + |A_2 \cap A_3 \cap A_4 \cap A_5| + |A_2 \cap A_3 \cap A_4 \cap A_6|$$
$$+|A_2 \cap A_3 \cap A_5 \cap A_6| + |A_2 \cap A_4 \cap A_5 \cap A_6| + |A_3 \cap A_4 \cap A_5 \cap A_6|)$$
$$-10(|A_1 \cap A_2 \cap A_3 \cap A_4 \cap A_5| + |A_1 \cap A_2 \cap A_3 \cap A_4 \cap A_6|$$
$$+|A_1 \cap A_2 \cap A_3 \cap A_5 \cap A_6| + |A_1 \cap A_2 \cap A_4 \cap A_5 \cap A_6|$$
$$+|A_1 \cap A_3 \cap A_4 \cap A_5 \cap A_6| + |A_2 \cap A_3 \cap A_4 \cap A_5 \cap A_6|)$$
$$+15|A_1 \cap A_2 \cap A_3 \cap A_4 \cap A_5 \cap A_6|$$
$$= {}^6C_2(6-2)! - 3 {}^6C_3(6-3)! + 6 {}^6C_4(6-4)! - 10 {}^6C_5(6-5)! + 15 {}^6C_6(6-6)!$$
$$= 360 - 3 \times 120 + 6 \times 30 - 10 \times 6 + 15 = 135.$$

Of course, if you have been thinking "derangements", life is a lot easier.

20. If two of the numbers are fixed, we can choose the remaining number in p_{n-2} ways. Since there are $^nC_2$ ways of choosing the two numbers to be fixed, there are $^nC_2 p_{n-2}$ permutations of n numbers which fix exactly two numbers.

   Is the answer to Exercise 19 $^6C_2 p_4$?

21. $^nC_k p_{n-k}$.

22. This is probably the most complicated problem in this chapter so I'm going to have a look at the cases $n = 2$, 3 and then give you another chance to prove the general result.

(a)
1	2	3	4
1		2	
	1		2

(b)
1	2	3	4	5	6
			2		
1					
	1			2	
					2

(c)
1	2	3	4	5	6
1	3		2	4	
1		3	2		4

Let a permutation have property $P(i)$ if i and $i+1$ are n apart. We have shown the start of such permutations in (a) and (b) above. In both cases $i = 1$.

The first row of (a) shows the permutation in which, so far, 1 goes to 1 and 3 goes to 2. Further 1 and 2 are $n = 2$ places apart so this permutation will have property $P(1)$.

The first row of (b) gives a permutation which, so far, has 1 going to 1 and 4 going to 2. No matter what we do with the other numbers, this permutation will have property $P(1)$.

Note that the permutations in (c) have property $P(1)$ and $P(3)$, since both 1 and 2 and 3 and 4 are $n = 3$ apart in the permutation.

Let's now try to count $P(1)$ in the case $n = 2$. We have shown two places 1 can go. However, we can put the 1 in the 3 or 4 column. So in fact 1 can be placed in 4 $(= 2n)$ positions provided 2 is **put in** the right place a distance of 2 $(= n)$ away.

So we have fixed the position of 1 and 2. We now have 2 positions in the permutation to fill. This can be done in 2! ways.

So $P(1) = 4 \cdot 2! = 8$.

In the case $n = 3$ we can work from (b) to get $P(1) = 6 \cdot 4! = 144$. Here 1 has 6 places it can go. This fixes the position of 2. We then have 4 numbers for 4 positions to give 4! more possibilities.

For $n = 3, P(i) = 6 \cdot 4!$ for $i = 1, 2, \ldots, 5$. So the number of permutations with property $P(i)$ is $\sum_{i=1}^{5} P(i) = 5 \cdot 6 \cdot 4! = 6!$. Note though that this drastically over counts these permutations. But that was what the principle of inclusion–exclusion was designed to do.

Now look at (c) and we'll try to count $P(1) \cap P(3)$. We know there are 6 choices for 1. These force the position of 2. The number 3 can then go in 4 ($= 2n - 2$) positions. These all force the position of 4. Then there are two numbers left for 5 and 6. There are 2 choices for these positions. Hence $P(1) \cap P(3) = 6 \cdot 4 \cdot 2!$.

As part of the inclusion–exclusion argument, we have to count $P(1) \cap P(3)$, $P(1) \cap P(4), P(1) \cap P(5), P(2) \cap P(4), P(2) \cap P(5)$. (We don't count $P(4) \cap P(5)$ since all permutations with property $P(4)$ fixed 5 and since the permutation is only six long, there is no room left for 5 to fix 6.) There are 6 types of permutation with the property $P(i) \cap P(j)$.

So for $n = 3$, the count for the number of permutations with property P is

$$6! - 6 \cdot 6 \cdot 4 \cdot 2! + \cdots .$$

The proportion of all permutations of 6 objects which have property P is

$$\frac{6!}{6!} - \frac{6 \cdot 6 \cdot 4 \cdot 2!}{6!} + \cdots = 1 - \frac{2}{5} + \cdots .$$

Now if the terms are getting smaller the proportion of permutations with property P is greater than $1 - \frac{2}{5} = \frac{3}{5} > \frac{1}{2}$. This tells us that more permutations have property P.

The solution for n in general is given later.

23. (i) $^{15}C_1 = 15$; (ii) $^7C_2 = 21$; (iii) $^{23}C_3 = 1771$.

24. (i) The number of solutions in positive integers equals the number of non-negative integer solutions minus the number of zero solutions $= 15 - 2 = 13$;

(ii) $21 - [3(^6C_1) + 3] = 0$ is obviously wrong. So we can't just subtract the number of solutions with one variable zero plus the number of solutions with two variables zero from the total number of non-negative solutions. This is because we've already counted the solutions with two variables zero, twice in the count of solutions with one variable zero. So our required number is $21 - [3(^6C_1) - 3] = 6$;

(iii) $1771 - [\text{one zero} - \text{two zeros} + \text{three zeros}]$
$= 1771 - [4(^{22}C_2) - 6(^{21}C_1) + 4] = 969$.

25. (i) $^{19}C_2 = 171$; (ii) $^{29}C_3$.

26. $^{n-1}C_{m-1}$.

27. Subtract 5 from 20. The number of solutions of the required type is the number of solutions of $x + y + z = 15$ in positive integers. This is $^{14}C_2 = 91$.

28. Subtract 5 from each of x, y and z and determine the number of positive integer solutions of $x + y + z = 5$. This is $^6C_1 = 6$.

29. Number = all positive integer solutions − [those with $x \geq 10 +$ those with $y \geq 10 +$ those with $z \geq 10$) − (those with $x \geq 10, y \geq 10 +$ those with $x \geq 10, z \geq 10 +$ those with $y \geq 10, z \geq 10$) + (those with $x \geq 10 \, y \geq 10, z \geq 10$)] $= {}^{19}C_2 - [3({}^{10}C_2) - 3 \times 0 + 0] = 36$.

    (Those with $x \geq 10$ are the same as all positive integer solutions of $x + y + z = 11$. There are no positive integer solutions of $x + y + z = 20$ with $x \geq 10, y \geq 10$.)

30. Number = all non-negative solutions − [(those with $x \geq 9 +$ those with $y \geq 10$) − (those with $x \geq 9, y \geq 10$)] $= {}^{22}C_1 - [({}^{13}C_1 + {}^{12}C_1) - {}^3C_1] = 0$.
    Clearly the largest $x + y$ can be is 17.

31. Number = all solutions with $x \geq 4, z \geq 2$
    $\quad$ − those with $x \geq 4, z \geq 2, y \geq 16$
    $= {}^{21}C_2 - {}^5C_2 = 95$.

32. This is the same as the number of solutions of $x + y + z = 10$ with $0 \leq x \leq 7, 0 \leq y \leq 9$ and $z \geq 0$.
    Number = all non-negative solutions
    $\quad$ − [(those with $x \geq 8 +$ those with $y \geq 10$)
    $\quad$ − (those with $x \geq 8, y \geq 10$)]
    $= {}^{12}C_2 - [({}^4C_2 + {}^2C_2) - (0)] = 59$.

33. (a)

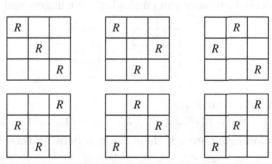

We'll assume that all these are different. I know that you can get the first by rotating the board to become the last, but from now on

we'll assume that the board is fixed in space and can't be moved. So the number is 6.

(b) We could do this again by trial and error and pretty pictures. The answer is 24.

(c) To get 40320 you'll have to be a little more systematic. If I said that $40320 = 8!$ would that help?

34. (a) The rooks have to occupy two columns. There are $^3C_2$ choices of columns. Then they also have to occupy two rows. There are $^3C_2$ choices of rows but you can choose which rook goes in which column. So there are $2^3C_2{}^3C_2 = 18$ ways.

(b) There are $^4C_3{}^4C_3$ ways of choosing a 3×3 array on the 4×4 board. We know that there are 3! ways of filling a 3×3 board with 3 non-taking rooks. So there are $3!(^4C_3)^2 = 96$ ways.

(c) For $n - 1$ non-taking rooks on an $n \times n$ board we have $(n - 1)!\ (^nC_{n-1})^2$ ways.

35. An $s \times s$ array can be chosen in $(^nC_s)^2$ ways. Each can be filled in $s!$ ways. So the answer is $s!(^nC_s)^2$.

36. $s!(^mC_s)(^nC_s)$.

37.   (i) $r_0 = 1, r_1 = 4, r_2 = 2!$. So $R(x, B) = 1 + 4x + 2x^2$;

   (ii) $1 + 9x + 18x^2 + 6x^3$;

   (iii) $1 + 16x + 72x^2 + 96x^3 + 24x^4$;

   (iv) $1 + 2x$;

   (v) $1 + 6x + 6x^2$;

   (vi) $1 + 12x + 36x^2 + 24x^3$;

   (vii) $1 + 20x + 120x^2 + 240x^3 + 120x^4$;

   (viii) $1 + 24x + 180x^2 + 360x^3$.

(Use Exercise 36 to save you employing your fingers and toes.)

38.

$R(x, D) = 1 + 7x + 10x^2 + 2x^3$.

39. $R(x, X) = 1 + 8x + 14x^2 + 4x^3$. (Note that placing two rooks in the same row is forbidden, even if there is a gap between these two rooks.)

$$R(x, Y) = 1 + 5x + 5x^2.$$

$$R(x, Z) = 1 + 7x + 16x^2 + 14x^3 + 4x^4.$$

40. $R(x, L) = (1 + x)^3$

    $R(x, M) = (1 + 2x)^2$

    $R(x, N) = (11 + x)(1 + 2x)(1 + 4x + 2x^2)$.

41. $R(x, B) = (1 + 4x + 2x^2)(1 + 9x + 18x^2 + 6x^3)$.

42. The answer is given in the next piece of text.

43. (a) $r_3(B) = r_3(B_1)r_0(B_2) + r_2(B_1)r_1(B_2)$

            $+ r_1(B_1)r_2(B_2) + r_0(B_1)r_3(B_2)$.

    (b) $r_4(B) = \sum_{i=0}^{4} r_{4-i}(B_1)r_i(B_2)$.

    (c) $r_k(B) = \sum_{i=0}^{k} r_{k-i}(B_1)r_i(B_2)$.

44. $R(x, B) = \sum_{k=0}^{n} r_k(B)x^k$

           $= \sum_{k=0}^{n} \sum_{i=0}^{k} r_{k-i}(B_1)r_i(B_2)x^k \sum_{t=0}^{n} r_s(B_1)x^s$

           $+ \sum_{s=0}^{n} r_t(B_2)x^t$

           $= R(x, B_1)R(x, B_2)$.

    (You may have to multiply out some of these products to convince yourself that what I've written is correct. Start with $n = 2$ and $n = 3$ and work up.)

45. Now $x \cdot R(x, D_1) + R(x, D_2) = (x + 4x^2 + 2x^3) + (1 + 6x + 6x^2) = 1 + 7x + 10x^2 + 2x^3 = R(x, D)$.

    A plausible, say, a very convincing argument is coming up in the next piece of text.

46. You should find that $R(x, X) = x \cdot R(x, X_1) + R(x, X_2)$.

47. Something like the equation of the last exercise holds.

48. This is Property II, you're after. It's coming up in the text ahead.

49. (a) If we put two non-taking rooks on B, then either one occupies square t or it doesn't. If it does, we can still put one non-taking rook on B_1. This can be done in $r_1(B_1)$ ways.

        If square t is not used, then we need to put two non-taking rooks on B_2. This is possible in $r_2(B_2)$ ways.

        Hence $r_2(B) = r_1(B_1) + r_2(B_2)$.

    (b) It's the same basic argument as in (a).

    (c) If square t is used then we put $k - 1$ rooks on B_1. If t is not used then we put k rooks on B_2. We find that

$$r_k(B) = r_{k-1}(B_1) + r_k(B_2).$$

    (This argument is only true for $1 \le k \le n$. What happens if $k = 0$?)

50. From Exercise 49(c), we have $r_k(B) = r_{k-1}(B_1) + r_k(B_2)$ for $1 \le k \le n$
and $r_0(B) = r_0(B_2)$.

So $R(x, B) = \sum_{k=0}^{n} r_k(B)x^k$
$$= [\sum_{k=1}^{n}(r_{k-1}(B_1) + r_k(B_2))x^k] + r_0(B_2)$$
$$= \sum_{k=1}^{n} r_{k-1}(B_1)x^k + \sum_{k=0}^{n} r_k(B_2)x^k$$
$$= x\sum_{k=1}^{n} r_{k-1}(B_1)x^k + R(x, B_2)$$
$$= x \cdot R(x, B_1) + R(x, B_2).$$

(To get the last equality let $k - 1 = I$. Then the summation goes from
$I = 0$ to $I = n - 1$. Note that if we can only put n non-taking rooks on
B then we can only put $n - 1$ non-taking rooks on B_1.)

51. We redo Exercise 37(vi) using Properties I and II. The other exercises
can be checked in a similar way.

Choose t to be the square shown. Then B_1 is a 2×3 board and B_2
is the 3×4 board with t deleted.

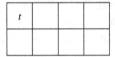

$R(x, B_1)$ can be found directly. Indeed, we've already worked that out
in part (v).

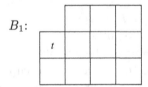

B_1:   Here B_1' is a 2×3 board.
$R(x, B_1') = 1 + 6x + 6x^2$. B_2' is shown.

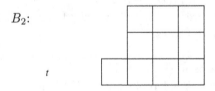

B_2: $R(x, B_1'') = 1 + 6x + 6x^2$.
$R(x, B_2'') = 1 + 9x + 18x^2 + 6x^3$.

Now $R(x, B_2') = x \cdot R(x, B_1'') + R(x, B_2'') = 1 + 10x + 24x^2 + 12x^3$.
Further $R(x, B_2) = x \cdot R(x, B_1') + R(x, B_2') = 1 + 11x + 30x^2 + 18x^3$.
Finally $R(x, B) = xR(x, B_1) + R(x, B_2) = 1 + 12x + 36x^2 + 24x^3$.

This is all a bit longwinded but, for larger boards, perhaps easier
than direct counting.

52. By the formula $p_3 = 2, p_5 = 44$ and $p_6 = 265$.

I'll do the calculation for p_6 only.

Now the board of forbidden positions is simply 6 non-interfering squares so its rook polynomial is
$$(1+x)^6 = 1 + 6x + 15x^2 + 20x^3 + 15x^4 + 6x^5 + x^6.$$

Hence
$$\sum_{k=0}^{n} (-1)^k (6-k)! r_k = 6! r_0 - 5! r_1 + 4! r_2 - 3! r_3 + 2! r_4 - 1! r_5 + 0! r_6$$
$$= 6! - 5!6 + 4!15 - 3!20 + 2!15 - 1!6 + 0!1$$
$$= 265.$$

53. The rook polynomial is $(1+x)^n = \sum_{i=0}^{n} {}^nC_i x^i$.

Hence $r_i = {}^nC_i$. So
$$\sum_{k=0}^{n} (-1)^k (n-k)! r_k = \sum_{k=0}^{n} (-1)^k (n-k)! {}^nC_k$$
$$= \sum_{k=0}^{n} (-1)^k \frac{(n-k)! n!}{k!(n-k)!}$$
$$= \sum_{k=0}^{n} (-1)^k \frac{n!}{k!}$$
$$= n! - n! + \frac{n!}{2!} - \frac{n!}{3!} + \cdots + (-1)^n \frac{n!}{n!}$$
$$= n! \left[1 - 1 + \frac{1}{2!} - \frac{1}{3!} + \cdots + (-1)^n \frac{1}{n!} \right].$$

This is precisely the formula for p_n described in Chapter 1.

54.

We note that the board is equivalent to the one shown below.

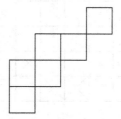

By Property I we get $(1+x)R(x,B) = (1+x)(1+5x+6x^2+x^3) = 1+6x+11x^2+7x^3+x^4$. So $r_0 = 1, r_1 = 6, r_2 = 11, r_3 = 7, r_4 = 1$.

Thus $\sum_{k=0}^{4} (-1)^k (4-k)! r_k = 4! - 3!6 + 2!11 - 1!7 + 0!1 = 4$.

(There are only the four possibilities: $A : c, B : t, C : f, D : s$; $A : s$, $B : c, C : t, D : f$; $A : s, B : c, C : f, D : t$; and $A : f, B : c, C : t, D : s$.)

55.

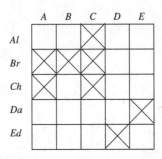

Property I gives the rook polynomial of the *bfp* as
$(1+x)(1+x)(1+6x+7x^2+x^3) = 1+8x+20x^2+21x^3+9x^4+x^5$.

So $r_0 = 1, r_1 = 8, r_2 = 20, r_3 = 21, r_4 = 9, r_5 = 1$.

Hence $\sum_{k=0}^{5} (-1)^k (5-k)! r_k = 5! - 4!8 + 3!20 - 2!21 + 1!9 - 0!1 = 14$.

56. Here $R(x,B) = (1+2x)(1+x)(1+2x)$ by Property I. So we get $R(x,B) = 1+5x+8x^2+4x^3$. If we put $r_0 = 1, r_1 = 5, r_2 = 8, r_3 = 4$, then

$$\sum_{k=0}^{3} (-1)^k (3-k)! r_k = 3! - 2!5 + 1!8 - 0!4 = 0.$$

This is clearly incorrect. After all the answer *must* be bigger than that of Exercise 54. What has gone wrong?

We answer that later.

57. All is revealed in the next part of the text.

58.

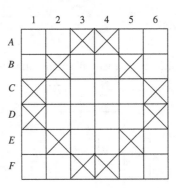

We show the board of forbidden positions.

This board is equivalent to three 2×2 boards put corner to corner. So by Property I, the rook polynomial is

$(1 + 4x + 2x^2)^3 = 1 + 12x + 54x^2 + 112x^3 + 108x^4 + 48x^5 + 8x^6$.

So $\sum_{k=0}^{6} (-1)^k (6-k)! r_k = 80$.

59. (a) Use Property II, standing at the left hand topmost square. This gives $R_n^m(x) = x R_{n-1}^{m-1}(x) + R(x, B_1)$, where B_1 is the $m \times n$ rectangle with the top left square deleted. Now use the square below the top left to give $R(x, B_1) = x R_{n-1}^{m-1}(x) + R(x, B_2)$. Continuing down this column, Property II gives $x R_{n-1}^{m-1}(x)$ at each stage and $R_{n-1}^{m-1}(x)$ as the last board when the m-th square in the first column goes.

So $R_n^m(x) = mx R_{n-1}^{m-1}(x) + R_{n-1}^m(x)$.

(Similarly we can find an expression for $R_n^m(x)$ in terms of $R_{n-1}^{m-1}(x)$ and $R_n^{m-1}(x)$.)

(b) Assume $m \leq n$. By Exercise 36,

$$kr_k(m, n) = k(k!)^m C_k{}^n C_k$$

$$= k^2(k-1)! \frac{m!}{k!(m-k)!} \frac{n!}{k!(n-k)!}$$

$$= \frac{(k-1)!m[(m-1)!]}{(k-1)!(m-k)!} \frac{n[(n-1)!]}{(k-1)!(n-k)!}$$

$$= mn[(k-1)!^{m-1}C_{k-1} {}^{n-1}C_{k-1}]$$

$$= mn\, r_{k-1}(m-1, n-1).$$

(Here $r_k(m, n)$ is the number of ways of placing k non-taking rooks on our $m \times n$ board.)

So

$$\frac{d}{dx}(R_n^m(x)) = \frac{d}{dx}\left(\sum_{k=0}^m r_k(m, n)x^k\right) = \sum_{k=1}^m kr_k(m, n)x^{k-1}$$

$$= \sum_{k=1}^m mn\, r_{k-1}(m-1, n-1)x^{k-1}$$

$$= mn \sum_{s=0}^{m-1} r_s(m-1, n-1)x^s = mn R_{n-1}^{m-1}(x).$$

Now

$$R_5^2(x) = R_4^2(x) + 2xR_4^1(x) = R_4^2(x) + 2x(1 + 4x)$$
$$= R_3^2(x) + 2xR_3^1(x) + 2x(1 + 4x) = R_3^2(x) + 2x(2 + 7x)$$
$$= R_2^2(x) + 2x(3 + 9x) = (1 + 4x + 2x^2) + 2x(3 + 9x)$$
$$= 1 + 10x + 20x^2.$$

60. Both of the boards below:

have rook polynomial $1 + 4x + 2x^2$. Does this generalise?

Chris Tuffley, a student in Year 12 from Christchurch, proved the following generalisation when he was living in Japan.

First we need a definition. Let $D(m, n, k, s)$ be the board below. This board is obtained by a block of size $m \times k$ from R_n^m and attaching it s units up the side of R_{n-k}^m.

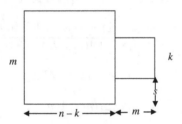

Tuffley's Theorem. $R(x, D(m, n, k, s)) = R_n^m(x)$ *for* $1 \le k \le m$ *and* $0 \le s \le m - k$.

Proof. The proof is by induction on k. Start with $s = 0$ and $k = 1$. By applying Property II successively to the squares on the bottom right of the board, we can show that

$$R(x, D(m, n, 1, 0)) = R_n^m(x).$$

By applying Property II in a similar way to $D(m, n, 1, s)$ we see that $R(x, D(m, n, 1, s)) = R(x, D(m, n, 1, 0))$.

Assuming the result is true for $k = r-1$, we can again apply Property II successively m times to the top row of the m by r block to give

$$R(x, D(m, n, r, s)) = mx R(x, D(m-1, n-1, r-1, s))$$
$$+ R(x, D(m, n-1, r-1, s)) \tag{1}$$
$$= mx R_{n-1}^{m-1}(x) + R_{n-1}^m(x) \quad \text{by induction}$$
$$= R_n^m(x) \text{ by Exercise 59(a).}$$

□

Chris adds the following comment. "There are some restrictions on the way in which we can take strips off and put them back on again. Otherwise you could eventually end up with a 1 by mn board, whose polynomial is $1 + mnx$. It may be that all strips being moved from vertical to horizontal have to be moved as a block".

Is Tuffley's Conjecture true?

61. (i) $S(4, 1) = 1$;
   (ii) $S(4, 3): a, b, cd; a, c, bd; a, d, bc; b, c, ad; b, d, ac; c, d, ab.$
       Hence $S(4, 3) = 6$;
   (iii) $S(4, 4) = 1$;
   (iv) $S(4, 0) = 0$ — you can't partition a set of size 4 into no subsets.

62. (i) $S(n, n) = 1$;
   (ii) $S(n, n-1)$; since there must be one subset of size two and the rest of size one, it depends on the number of ways there are of choosing two elements from n. So $S(n, n-1) = {}^n C_2$.

63. (a) $S(4, 2) = 7, S(3, 2) = 3$ and $S(3, 1) = 1$.
       So $S(4, 2) = 2S(3, 2) + S(3, 1) = S(3, 2) + 4S(3, 1)$.
       There are many other possibilities.
   (b) $S(5, 2) = 15$ (you'll need to check this out). $S(4, 2) = 7$ and $S(4, 1) = 1$. So $S(5, 2) = 2S(4, 2) + S(4, 1) = S(4, 2) + 8S(4, 1)$.
   (c) Could $S(n, 2) = 2S(n-1, 2) + S(n-1, 1)$? Read on.

64. (a) Let the set of size 5 be $\{a, b, c, d, e\}$. When we are dividing this set into three sets, then either $\{a\}$ is a set by itself or a is in some other set. In the former case we can divide the set $\{b, c, d, e\}$ into two sets in $S(4, 3)$ ways. In the latter case we have to put $\{b, c, d, e\}$ into three sets and then add back a in one of three ways. This can be done in $3S(4, 3)$ ways.
   (b) The argument is precisely the same as that in (a) except we replace three by r and two by $r - 1$, in each case.

65. $S(6,0) = 0, S(6,1) = 1, S(6,6) = 1.$

        $S(6,2) = 2S(5,2) + S(5,1) = 31.$

        $S(6,3) = 3S(5,3) + S(5,2)$

$$= 3[3S(4,3) + S(4,2)] + 15$$

$$= 3[3 \times 6 + 7] + 15 = 90.$$

        $S(6,4) = 4S(5,4) + S(5,3) = 4^5 C_2 + 25 = 65.$

        $S(6,5) = {}^6C_2 = 15$ (or $S(6,5) = 5S(5,5) + S(5,4) = 15$).

66. Now $S(3,2) = 3$. The six ways of putting the 3 balls in the two boxes is shown below.

Alternatively, we can divide the coloured balls into two lots in $S(3,2)$ ways. But because there are 2 different boxes, these "lots" can be put in the boxes in 2 ways per lot. Hence there are $2S(3,2) = 2!S(3,2)$ ways of assigning the coloured balls.

67. (a) An assignment of 4 objects into 3 sets can be done in $S(4,3)$ ways. Because the boxes are different, each set decomposition leads to 3! ways of assigning the balls. Hence the required number is $3!S(4,3) = 24.$

     (b) $n!S(m,n).$

68.

$$S(4,0) = \sum_{r=0}^{0} (-1)^r \frac{(0-r)^4}{r!(0-r)!} = 0.$$

$$S(4,1) = \sum_{r=0}^{1} (-1)^r \frac{(1-r)^4}{r!(1-r)!} = \frac{1^4}{0!1!} - \frac{0^4}{1!0!} = 1.$$

$$S(4,2) = \sum_{r=0}^{2} (-1)^r \frac{(2-r)^4}{r!(2-r)!} = \frac{2^4}{0!2!} - \frac{1^4}{1!1!} + 0 = 7.$$

$$S(4,3) = \frac{3^4}{0!3!} - \frac{2^4}{1!2!} + \frac{1^4}{2!1!} - 0 = 6.$$

$$S(4,4) = \frac{4^4}{0!4!} - \frac{3^4}{1!3!} + \frac{2^4}{2!2!} - \frac{1^4}{3!1!} + 0 = 1.$$

69. $nS(m-1,n) + S(m-1,n-1)$

$$= n\sum_{r=0}^{n} \frac{(-1)^r(n-r)^{m-1}}{r!(n-r)!} + \sum_{s=0}^{n-1} \frac{(-1)^s(n-1-s)^{m-1}}{s!(n-1-s)!}$$

$$= \frac{n^m}{n!} + \sum_{r=1}^{n} \frac{(-1)^r(n-r)^{m-1}n}{r!(n-r)!} + \sum_{s=0}^{n-1} \frac{(-1)^s(n-1-s)^{m-1}}{s!(n-1-s)!}$$

$$= \frac{n^m}{n!} + \sum_{s=0}^{n-1} \left[\frac{(-1)^{s+1}(n-1-s)^{m-1}n}{(s+1)!(n-1-s)!} + \frac{(-1)^s(n-1-s)^{m-1}}{s!(n-1-s)!} \right]$$

$$= \frac{n^m}{n!} + \sum_{s=0}^{n-1}(-1)^{s+1}\frac{(n-1-s)^{m-1}}{s(n-1-s)!}\left[\frac{n}{s+1} - 1 \right]$$

$$= \frac{n^m}{n!} + \sum_{s=0}^{n-1}(-1)^{s+1}\frac{(n-1-s)^m}{(s+1)!(n-1-s)!}$$

$$= \frac{n^m}{n!} + \sum_{t=1}^{n}(-1)^t\frac{(n-t)^m}{t!(n-t)!} = S(m,n).$$

70. (a) $S'(4,1)$ is minus the coefficient of x in $x(x-1)(x-2)(x-3)$. This is minus the constant term of $(x-1)(x-2)(x-3)$, which is 6.

$S'(5,2)$ is minus the coefficient of x^2 in $x(x-1)(x-2)(x-3)(x-4)$. This is 50.

$S'(6,2)$ is the coefficient of x^2 in $x(x-1)(x-2)(x-3)(x-4)(x-5)$. This is 274.

(b) $S'(m,m-1)$ is the magnitude of the coefficient of x^{m-1} in

$$(x-1)(x-2)\cdots(x-m+1).$$

This is the size of the coefficient of x^{m-2} in

$$(x-1)(x-2)\cdots(x-m+1)$$

which is

$$|-1+(-2)+\cdots+(-m+1)| = 1+2+\cdots+m-1 = \tfrac{1}{2}m(m-1).$$

(c) $x(x-1)\cdots(x-m+2) = S'(m-1,m-1)x^{m-2}$
$$-S'(m-1,m-2)x^{m-2} + S'(m-1,m-3)x^{m-3} - \cdots.$$

Multiply both sides of this equation by $x-m+1$.

Then

$x(x-1)\cdots(x-m+1)$
$= S'(m-1,m-1)x^m - S'(m-1,m-2)x^{m-1}$
$+S'(m-1,m-3)x^{m-2} - \cdots - (m-1)S'(m-1,m-1)x^{m-1}$
$+(m-1)S'(m-1,m-2)x^{m-1} - \cdots.$

But the left-hand side of this last equation is equal to
$S'(m,m)x^m - S'(m,m-1)x^{m-1} + S'(m,m-2)x^{m-2-\cdots}.$

Hence $S'(m, m) = S'(m - 1, m - 1) = (m - 1)S'(m - 1, m) + S'(m - 1, m - 1)$, since $S'(m - 1, m) = 0$.

$$S'(m, m - 1) = (m - 1)S'(m - 1, m - 1) + S'(m - 1, m - 2)$$

$$S'(m, m - 2) = (m - 1)S'(m - 1, m - 2) + S'(m - 1, m - 3), \quad \text{etc.}$$

So in general,

$$S'(m, n) = (m - 1)S'(m - 1, n) + S'(m - 1, n - 1).$$

71. I'll use numbers to represent the basketballers, with higher numbers representing taller players. I'll also assume that I have $2n$ players altogether and answer the questions as cases of this general situation.

If $n = 1$ there is only one possible photograph.

$n = 2:$ 3 4   2 4
           1 2   1 3

$n = 3:$ 4 5 6   3 5 6   3 4 6   2 5 6   2 4 6
        1 2 3   1 2 4   1 2 5   1 3 4   1 3 5

$n = 4:$ 5 6 7 8   4 6 7 8   4 5 7 8   4 5 6 8   3 6 7 8
        1 2 3 4   1 2 3 5   1 2 3 6   1 2 3 7   1 2 4 5
        3 5 7 8   3 5 6 8   3 4 7 8   3 4 6 8   2 6 7 8
        1 2 4 6   1 2 4 7   1 2 5 6   1 2 5 7   1 3 4 5
        2 5 7 8   2 5 6 8   2 4 7 8   2 4 6 8
        1 3 4 6   1 3 4 7   1 3 5 6   1 3 5 7

$n = 5:$ You should have found 42 in this case but the numbers of possibilities here are beginning to be a little too large to put into print.

72. I'll stay away from trying to generalise at the moment. Do the numbers look like anything familiar though?

73. It's probably too soon to find a formula here but if you have experimented with small cases you would have found that the numbers, 1, 2, 5, 14, and 42 came up. You might expect that there is some relation between the Change Problem and the Photo Problem.

74. Again, small cases lead to the same numbers.

75. It pays to do counting like this systematically. So I'll start with a string of 4 left brackets, then 3, then 2 and so on. So I get:

     (((( ))))   ((( ) ( )))   ((( )) ( ))   ((( ))) ( )
     (( ) (( )))   (( ) ( ) ( ))   (( )) (( ))   (( )) ( ) ( )
     (( ) ( )) ( )   ( ) ((( )))   ( ) (( ) ( ))   ( ) (( )) ( )
     ( ) ( ) (( ))   ( ) ( ) ( ) ( )

76. Some of these can be found in the text and in Exercise 80. Can you find an equivalence between the Triangulation Problem and the Change Problem?

77. The first two equalities are a direct use of the definition of the Binomial Coefficients. The last result is a known Binomial Coefficient identity. However, you can prove it by first getting a generalisation. Expanding $(1 + x)^a(1 + x)^{b-a} = (1 + x)^b$, and comparing coefficients gives Vandermonde's Identity, $\sum_i {}^aC_i \, {}^{b-a}C_{t-i} = {}^bC_t$. If we now let $b = 2n$ and $a = n$, the last equation follows.

78. Equation (2) is a direct use of the definition of the Binomial Coefficients. Can you prove that equation (3) holds?

79. The Triangulation Problem is essentially about finding the number of triangulations of a given sized polygon with one side of the polygon given a special role. Call the special edge the **base** of the polygon. Now the base must be one side of a triangle in the triangulation — a base triangle. Essentially there is another polygon to the left of this base triangle that uses an edge of the triangle. The same thing occurs on the right. If I use Mathematical Induction, I can assume that each of these smaller polygons can be triangulated in a Catalan number of ways. Multiplying these two smaller Catalan numbers together gives the number of triangulations where the base has a given base triangle. I can now sum these products over all base triangles to give the number of triangulations of the original polygon. Using equation (3) with the Mathematical Induction, I can prove that the Catalan numbers do count the number of triangulations of any polygon.

80. Let me first illustrate this by doing the case $n = 8$. Take the photo arrangement

$$4 \; 6 \; 7 \; 8$$

$$1 \; 2 \; 3 \; 5$$

Now put left brackets in positions 1, 2, 3 and 5 of the brackets and put right brackets in positions 4, 6, 7 and 8. Then we are mapping the photo to the bracket arrangement $((()()))$. This map will only cause problems if, at some stage along the bracket arrangement, the map sends more right brackets than left ones. But this would mean that there is a sequence of numbered people in the photograph like the one below

$$b_1 b_2 \cdots b_{s-1} b_s$$

$$a_1 a_2 \cdots a_r$$

where $s > r, a_i < a_{i+1}, a_i < b_i$ and $b_j < b_{j+1}$. Then $a_{r+1} \geq b_s$, and so the photograph would not have been legitimate. So every photo gets sent to a unique bracket by this map.

Now we can also map the brackets to the photos by the reverse of the map above. A similar argument to the one above shows that the image of every bracket is a photo. Hence the photos and brackets are equi-numerous.

This argument works for every value of n.

Of course, as the result of this and earlier work we now know that the Photo Problem and the Change Problem are also equi-numerous. However, you might like to show this directly.

81. You will find a number of applications on the web if you search for "Catalan numbers". One of these occurrences of the Catalan numbers is in a developing area of mathematics called Permutation Patterns. Here researchers are interested in how many permutations of n objects avoid a particular sub-permutation. For instance, how many permutations avoid 231? Now I should point out that this doesn't mean that only 231 as a block is to be avoided (as in the permutation 7623185), but also anywhere in order (as in 7263185) and in kind (as in 7612385, where 685 is considered to be a "231" as it first rises and then falls below the first member of the sub-permutation). You might like to try to prove that the number of permutations on n objects that is 231-avoiding is L_n.

Note that each such permutation can be written as some permutation before n and some permutation after. Here the something before and something after has to be 231-avoiding. Further every term before is less than every term after. Then equation (3) can be used to justify the Catalan count.

22. (Cont'd): Look at the general case. The number of permutations with property $P(i)$ is $2n[(2n-2)!]$. There are $2n-1$ possibilities for i so the number of permutations with property $P(i)$ for $i = 1, 2, \ldots, 2n-1$ is overcounted at $2n[(2n-2)!](2n-1) = (2n)!$.

The number of permutations with property $P(i) \cap P(j)$ is

$$(2n)(2n-2)[(2n-4)!].$$

How many terms are there of the form $P(i) \cap P(j)$? If $i = 1$ or $2n-1$, then there are $2n-3$ possible j's it can be paired with. Otherwise i can be paired with $1, 2, \ldots, i-2, i+2, \ldots, 2n-1$. There are $2n-4$ possibilities here. Hence there are $\frac{1}{2}[2(2n-3) + (2n-3)(2n-4)] = (2n-3)(n-1)$ terms of the form $P(i) \cap P(j)$.

Hence the number of permutations with property P is

$$(2n)! - (2n-3)(n-1)[(2n-4)!(2n)(2n-2)] + \cdots.$$

The proportion with property P is

$$\frac{(2n)!}{(2n)!} - \frac{(2n-3)(n-1)[(2n-4)!(2n)(2n-2)]}{(2n)!} + \cdots = 1 - \left(\frac{n-1}{2n-1}\right) + \cdots .$$

Now if the number of terms with the alternating signs is decreasing in magnitude, the proportion is greater than

$$1 - \left(\frac{n-1}{2n-1}\right) = \frac{n-1}{2n-1} > \frac{n-\frac{1}{2}}{2n-1} > \frac{1}{2}.$$

This is the required result.

(I will leave you to show that the number of permutations with properties $P(i_1), P(i_2), P(i_3)$ is greater than those with $P(i_1) \cap P(i_2)$, $P(i_1) \cap P(i_4)$ and so on.)

56. (cont'd) The problem is that we have a 4×4 board and so we should have taken $\sum_{k=0}^{4} (-1)^k (4-k)! r_k$. However, this requires us to have an r_4. But what can this be? After all, the rook polynomial is $1 + 5x + 8x^2 + 4x^3$ and it has no x^4 term. Well then r_4 must be zero.

Hence $\sum_{k=0}^{4} (-1)^k (4-k)! r_k = 6$.

Chapter 7

Creating Problems

7.1. Introduction

In this chapter I want to give you a chance to generate your own problems. This is partly because it is fun, partly because it will reinforce connections you have across various parts of the subject, partly because it will increase your problem solving ability, and partly because in the process you will learn some interesting mathematics.

What I intend to do here is to look at the humble square and see where I can get from there by asking questions (not always my own questions). These questions are, hopefully, the kind of thing a mathematician might ask if they had come across the square for the first time. So they will not only be questions that explore the square itself, but they will also look at potential connections with other areas of mathematics to see if the square can generalise or extend things that can be done in other areas. So I'll look at counting, packing, and intersecting, at chessboards and at an extension of trigonometry that I call squigonometry.

Mathematics is all about asking the right questions. If you chew your problems down into smaller and smaller pieces by the use of the right questions, you'll find them so much easier to solve. Whether or not I have asked the right questions here may not be clear. However, I have tried to question to see what I can find out.

One final comment, there are now only three differences between you, in your work with mathematical problems, and a professional mathematician. Largely the actual techniques you use mathematicians use too.

The first difference is that the problems mathematicians work out have not been solved — they are open questions. Your problems are usually posed by someone who knows they have an answer and what the answer is. (I should point out though that this is not necessarily the case here. I don't know all the answers in this chapter — nor in other chapters either.) The second difference is that mathematicians have more experience and knowledge to draw upon but you'll catch up. The final difference is that your brains are more flexible, agile and adept than mathematicians. You'll quickly catch up with us. In fact, you may already be solving some problems faster than we can.

Happy problem creating.

7.2. Counting

Before I get too involved with things I'd better make sure I know what a square is. Obviously it's a four sided figure with all sides equal and with the angle between each pair of adjacent sides equal to 90°.

Exercises

1. Correct the error in the definition of a square given above.
   Can you reduce the amount of information given in your definition and still retain a correct definition of a square?
2. The following figure represents a cube *ABCDEFGH* with sides of length 4 cm. The points *X, Y, U, V* are on the edges *FG, CG, AE, AD*, respectively, with $AU = AV = GX = GY = 3$ cm. Is *XUVY* a square?

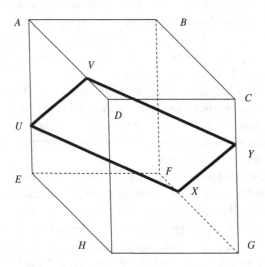

(Scottish Mathematical Council — Mathematical Challenge, 1988–1989, problem III.4.)

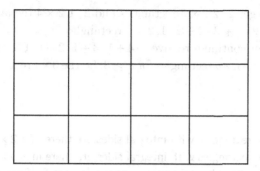

Figure 7.1.

Now in Chapter 5 of *First Step*, I asked two questions which I think we should look at again and try to generalise. In Exercise 6 of Chapter 5, an $n \times n$ square was divided into n^2 unit squares (that is, squares of side 1). The problem was to find out how many squares there are in this $n \times n$ square. Later, in Exercise 14, I divided a 6×9 rectangle into unit squares and again asked how many squares there were in the 6×9 rectangle. Clearly we can generalise the 6×9 case to the $m \times n$ case. The answer to all these three questions can be found in Chapter 5.

Example 1. How many rectangles are there in the 3×4 rectangle in Figure 7.1?

Before you do this question I should say that I want to count **all** the $m \times n$ rectangles that appear. So let's start systematically. How many 1×1 rectangles are there? In any row there are four of these. There are three rows so we have twelve 1×1 rectangles altogether.

How many 1×2 rectangles are there? In each row there are three such rectangles. So there are nine 1×2 rectangles.

Similarly the number of 1×3 rectangles is $3 \cdot 2 = 6$ and the number of 1×4 rectangles is $3 \cdot 1 = 3$.

So we have $3 \cdot 4 + 3 \cdot 3 + 3 \cdot 2 + 3 \cdot 1$ rectangles of the form $1 \times b$ where $b = 1, 2, 3, 4$.

Now look at the $2 \times b$ rectangles. Using the first two rows we get four 2×1 rectangles. Using the second two rows we get another four. So there are $2 \cdot 4$ 2×1 rectangles.

Using the first two rows we get three 2×2 rectangles. Using the last two rows we get another three. So there are $2 \cdot 3$ 2×2 rectangles.

Similarly we get $2 \cdot 2$ 2×3 rectangles and $2 \cdot 1$ 2×4 rectangles, to give a total of $2 \cdot 4 + 2 \cdot 3 + 2 \cdot 2 + 2 \cdot 1, 2 \times b$ rectangles.

The argument continues to give $1 \cdot 4 + 1 \cdot 3 + 1 \cdot 2 + 1 \cdot 1$ $3 \times b$ rectangles. Clearly there are no $a \times b$ rectangles for $a \geq 3$. So the 3×4 rectangle contains 60 rectangles.

Exercises

3. (a) How many rectangles with integral sides are there in a 2×3 rectangle?
   (b) How many rectangles with integral sides are there in a $p \times q$ rectangle?
4. Now that you've seen how simple an answer you get for Exercise 3(b), give a two line proof.
5. (a) How many 4×5 rectangles are there in a 7×9 rectangle?
   (b) In a $p \times q$ rectangle how many $a \times b$ rectangles are there, where $a \leq p$ and $b \leq q$.
6. For what $p \times q$ rectangles is the number of $a \times b$ rectangles with $a \leq b$, equal to the number of $b \times a$ rectangles (i) for a particular value of a and b; (ii) for all possible values of a and b, $1 \leq a \leq p$, $1 \leq b \leq q$?
7. We say that an $a \times b$ rectangle is **congruent** to an $a' \times b'$ rectangle if $a = a'$ and $b = b'$ or if $a = b'$ and $b = a'$. Two rectangles are said to be of a **different type** if they are not congruent.
   How many different types of rectangles are there in a $p \times q$ rectangle? (Assume $p \leq q$ without loss of generality.)

Already in Chapter 5 (Exercise 26) we extended from squares to equilateral triangles. Presumably we can do the same thing for the sort of questions we've been asking above. And, of course, there's no need to confine our attention to the plane.

Exercises

8. Consider the equilateral triangle below, whose sides are of length 3.

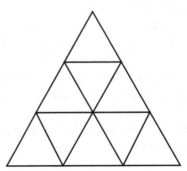

    (a) How many rhombuses are there?

    (b) How many trapezia are there?

    (c) Generalise.

9. Divide a cube whose edges are of length 4 into smaller cubes with edges of size 1.

   How many cuboids are there inside this cube?

   Generalise.

10. Can a regular tetrahedron whose edges are of length n be divided up into tetrahedra of unit edge length? If so, how many tetrahedra are there in such a tetrahedron?

This move into three dimensions reminds me of other cube-type problems. A couple of them follow.

Exercises

11. A $3 \times 3 \times 3$ cube is painted red and is then cut into 27 unit cubes. How many of these small cubes will have paint on **exactly** 2 faces?

    ("Problems, Problems, Problems" Vol. 2, Canadian Mathematics Competition, 1989, Question 2, page 39.)

12. A $5 \times 5 \times 5$ cube is formed using $1 \times 1 \times 1$ cubes. A number of smaller cubes are removed by punching out the 15 designated columns from front to back, top to bottom and side to side. How many smaller cubes are left? (Canadian Mathematics Competition, Gauss Contest, 1990, Question 5.)

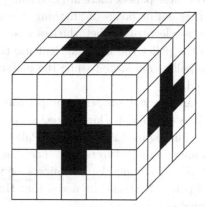

13. Rectangles, squares and equilateral triangles can each be packed together to fill the plane. What regular solids can be packed together to fill three dimensional space? How many smaller ones fit together to make a larger one?

14. Generalise all of the above and extend to completely new problems.

7.3. Packing

It's pretty clear that we can always fit 1×1 rectangles exactly into any $p \times q$ rectangle. There's never any space left over. But what about 1×2 rectangles in $p \times q$ rectangles? Is the fit exact?

Exercises

15. Into which of the following rectangles can 1×2 rectangles be fitted exactly?
    (i) 4×6;   (ii) 4×5;   (iii) 5×5;   (iv) 3×7;   (v) 5×7;   (vi) 3×21.
    Where the fit is not exact, how much of the larger rectangle is left over?

16. Conjecture the amount left over if we fit 1×2 rectangles into a $p \times q$ rectangle in the best possible way.
    Prove your conjecture.

17. Find the smallest amount uncovered when 1×3 rectangles are packed into the following larger rectangles.
    (i) 4×4;   (ii) 4×5;   (iii) 4×6.

18. Conjecture the amount left over if we fit 1×3 rectangles into a $p \times q$ rectangle in the best possible way.
    Prove your conjecture.

19. Repeat the last exercise with 1×4 rectangles in a $p \times q$ rectangle.

What's the point of doing this anyway? One reason is pallets. These are the things that forklifts pick up and move around all day long in warehouses. And perched on top of these pallets there are frequently rectangular shaped things, like boxes of tissues or boxes of anything. Clearly the more boxes that can be fitted on each level of a pallet, the less waste space on the pallet. The less wastage on the pallet the more efficient use there is of the whole warehouse when the pallets are stored while the material is waiting to be shipped away and sold.

So that pallets are packed properly, proprietors provide plans for producing parsimonious practical packing of parcels of products on pallets. One such plan is shown in Figure 7.2. It is from a Palletising Chart prepared by the Australian Paper Manufacturers Ltd. The diagram gives the best packing (least wastage) of rectangular containers of various sizes on an Australian Standard pallet. It should be noted that these "best packings" are not necessarily unique.

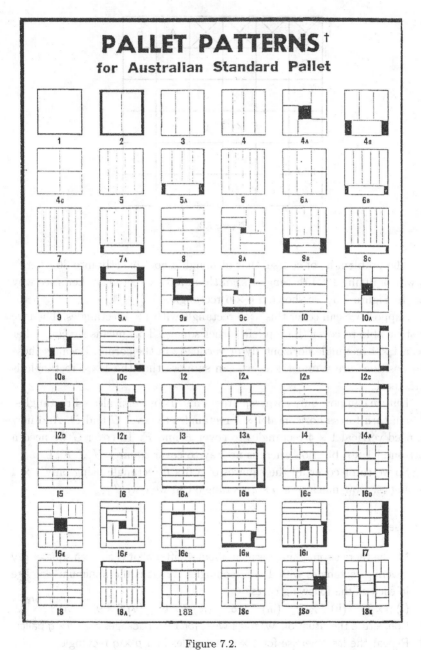

Figure 7.2.

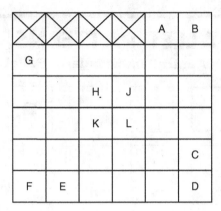

Figure 7.3.

The problem is, of course, that although 4 goes exactly into 36, there's no way of fitting 1×4 rectangles *exactly* in a 6×6 rectangle. Let's see why not. In Figure 7.3 I've shown a 6×6 rectangle.

Suppose we can cover this 6×6 rectangle with 1×4 rectangles. Then we must be able to cover the top left corner square. This means we can either put a 1×4 rectangle horizontally or vertically in the corner. Because these two positions are symmetrical, we can suppose that the 1×4 rectangle is horizontal.

But we can now only cover squares A and B by vertical 1×4 rectangles. This in turn forces horizontal 1×4 rectangles to cover C and D. And now we need vertical 1×4 rectangles to cover E and F. To cover G we need a horizontal 1×4 but now there is no way to cover squares H, J, K, L.

So we can't cover all the squares of a 6×6 rectangle using only 1×4 rectangles. How much of a $p \times q$ rectangle can we cover?

Exercises

20. Maybe there's something peculiar about 4, after all it is an even square. Try covering rectangles with 1×5 rectangles. What is the minimum wastage in each of the following cases?
    (i) 6×7;   (ii) 7×7;   (iii) 7×8;   (iv) 8×8;   (v) 8×9;   (vi) 12×13.
21. Conjecture the situation, wastage-wise, for 1×5 packages on a $p \times q$ pallet.
22. Repeat the last exercise for $1 \times n$ rectangles in a $p \times q$ rectangle.
23. (a) What's the situation for a $2 \times n$ rectangle in a $p \times q$ rectangle?
    (b) How about 3×3 squares in $p \times q$ rectangles?

24. Now in everything we've done so far we've lined up the sides of the smaller rectangles so that they were parallel to the sides of the larger rectangle. Can wastage be reduced by twisting the smaller rectangles?
25. How much space is left over when $l \times m \times n$ bricks are placed in a $p \times q \times r$ box?

There are good practical reasons for sticking to the case where the sides of the smaller rectangles are parallel to the sides of the larger rectangles. When sheets of paper, for example, are being cut from rolls of paper, the machinery is most easily set up to make cuts along and across the rolls. So this case has many practical applications. Initially it is surprising that it is more economical to have squares scattered about at angles. That certainly is what happens when the smaller rectangle is quite a lot smaller than the larger one (see "On Packing Squares with Equal Squares" by P. Erdös and R. L. Graham, Journal of Combinatorial Theory A, 1975 or "Mathematics Today", ed. L A Steen, Springer-Verlag, New York, 1978, page 209.) But it's also true in less extreme situations such as the one in Figure 7.4. This picture shows the best way to pack squares of side length 2 into a square of side length 5.

Now we don't always want to fit rectangles into rectangles. Suppose you are mass producing dresses. You will want to cut the pieces from the roll of cloth in the most efficient way possible. You will want to throw away as little material as possible. No one knows the absolute best method of doing this. Although there are computer programs which work fairly well, no one can prove they do the best possible job. It turns out that there are many people who get paid a great deal of money to minimise the wastage of dress and other material.

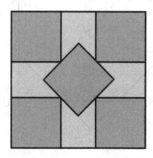

Figure 7.4.

Of course the other problem with dresses is that the pattern has to go the right way. And imagine pin striped suits with the stripe going sometimes horizontally and sometimes vertically. (Did you see that episode of MASH?)

7.4. Intersecting

Look at Figure 7.5. It shows two squares with the same centre O, but the axes of one are at $45°$ to the axes of the other. Suppose the side length of the square $ABCD$ is t and the side length of the square $A'B'C'D'$ is T.

There's a whole series of questions that we can ask about the octagon $PQ'QR'RS'SP'$ or the triangles APP', BQQ', $A'PQ'$, etc.

Exercises

26. (i) Find the smallest value of T in terms of t if $\triangle APP'$ has zero area;
    (ii) Find the largest value of T in terms of t if $\triangle A'PQ'$ has zero area.
27. (a) It seems intuitively clear that $\triangle$'s APP', BQQ', CRR', DSS' have equal areas. Prove it.
    (b) It seems intuitively clear that $\triangle$'s $A'PQ', B'QR', C'RS', D'SP'$ have equal areas. Prove it.
    (c) Find T in terms of t if $\triangle APP'$ has the same area as $\triangle A'PQ'$.

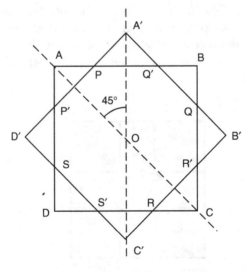

Figure 7.5.

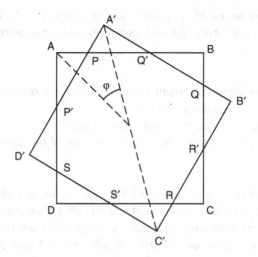

Figure 7.6.

28. Find T in terms of t if the sum of the areas of the triangles APP', BQQ', CRR', DSS' is equal to the area of the octagon $PQ'QR'RS'SP'$.

Now rotate the square $A'B'C'D'$ so that its axes are at an angle φ to the axes of the square $ABCD$. This is shown in Figure 7.6.

Exercises

29. Let φ be fixed.
    (i) Find the smallest value of T in terms of t for which $\triangle APP'$ has zero area.
    (ii) Find the largest value of T in terms of t for which $\triangle A'PQ'$ has zero area.
    Use a continuity argument to show that there is some value of T for which the area of $\triangle APP'$ is equal to the area of $\triangle A'PQ'$.
30. Repeat Exercise 27 on the situation of Figure 7.6.
31. Repeat Exercise 28 on the situation of Figure 7.6.
32. (a) What is the smallest value of T, such that for all values of φ, $ABCD$ is always totally ***inside*** $A'B'C'D'$?
    (b) What is the largest value of T, such that for all values of φ, $ABCD$ is always totally ***outside*** $A'B'C'D'$?

If we knew where the centre of an equilateral triangle was, then we could go through Exercises 26 to 28 using equilateral triangles. Well there are various things we could use. After all, in Chapter 2 we learnt about incentres and circumcentres and even centroids and orthocentres.

Exercises

33. Show that the incentre, circumcentre, centroid and orthocentre are the same point in an equilateral triangle.
    (If you've forgotten what these are, see Chapter 2.)
34. Are there any other triangles for which some two of the above four points are the same?

Since the centre of an equilateral triangle seems well defined (after Exercise 33) we can look at two equilateral triangles with the same centre. Figure 7.7 is one possible arrangement. The angle between the axes of the two triangles is 60°. Suppose $\triangle PQR$ has side length t and $\triangle P'Q'R'$ has side length T.

Exercises

35. (i) Find the smallest value of T for which $\triangle PA'B$ has zero area.
    (ii) Find the largest value of T for which $\triangle P'BB'$ has zero area.
    Is it possible for the areas of $\triangle PA'B$ and $\triangle P'BB'$ to be equal?
36. Find T in terms of t if $AA'BB'CC'$ is a regular hexagon.
37. Find T in terms of t if the sum of the areas of $\triangle$'s $PA'B, QB'C, RC'A$ is equal to the area of the hexagon $AA'BB'CC'$.

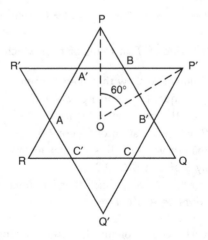

Figure 7.7.

38. Let the angle POP' become φ and stay fixed. Find T in terms of t if the area of $\triangle A'PB$ is equal to the area of $\triangle BP'B'$.

39. (a) What is the smallest value of T, such that for all values of φ, $\triangle PQR$ is always totally **inside** $\triangle P'Q'R'$?

    (b) What is the largest value of T, such that for all values of φ, $\triangle PQR$ is always totally **outside** $\triangle P'Q'R'$?

40. Now you've got the idea, play with two concentric (with the same centre) rectangles.

    Devise your own set of questions.

    Repeat the game using regular pentagons (is that hard?), regular hexagons, regular n-gons.

But there's no reason why we should stick to concentric figures of the same shape. Have a look at Figure 7.8. There we have an equilateral triangle and a square. Suppose the triangle has side length e and the square has side length s.

Exercise

41. (a) When is the area of $\triangle PL'M$ zero?

    (b) When is the area of $\triangle ALL'$ zero?

    (c) Is it possible for the areas of $\triangle$'s $PL'M, ALL'$ to be equal?

    (d) If the answer to (c) is "yes", find the relationship between e and s for it to happen.

    (e) Can area $\triangle PL'M =$ area $CDNN'$? If so, when?

    (f) Can area $\triangle ALL' =$ area $CDNN'$? If so, when?

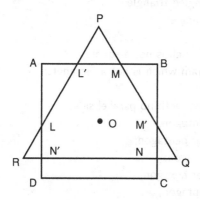

Figure 7.8.

(g) When area $\triangle PL'M =$ zero, what is the area of $\triangle QNM'$?

(h) If area $\triangle QNM'$ is zero, is P inside the square?

(i) What is the smallest value of e, such that, if $\triangle PQR$ is rotated through $360°$, $\triangle PQR$ always lies totally **outside** the square?

(j) What is the largest value of e, such that, if $\triangle PQR$ is rotated through $360°$, $\triangle PQR$ always lies totally **inside** the square?

(k) Any other questions?

All those questions and at no time did we leave the plane. Clearly it's time to get stuck into 3-dimensions. Before we do that though, let's go back to Figure 7.4. We could think of the square $A'B'C'D'$ as cutting off the corners of square $ABCD$. The result is a truncated, or dismembered in the nicest possible way, square. It's clear that truncating $ABCD$ with too large a square, has no affect on $ABCD$. But we can also truncate $ABCD$ to give an octagon, a regular octagon or a square (if $A'B'C'D'$ is small enough). Let's now try truncating the odd cube.

Exercise

42. Take one wooden cube and a saw. Use the saw to make a cut along any plane in space. Can you cut the cube so that each of the following faces are formed?

    (i) an equilateral triangle;

    (ii) an isosceles triangle;

    (iii) a right angled triangle;

    (iv) an acute angled triangle;

    (v) an obtuse angled triangle;

    (vi) a scalene triangle;

    (vii) a square;

    (viii) a rectangle which is not a square;

    (ix) a parallelogram which is not a rectangle;

    (x) a trapezium;

    (xi) a quadrilateral with no parallel sides;

    (xii) a regular pentagon;

    (xiii) a non-regular pentagon;

    (xiv) a regular hexagon;

    (xv) a non-regular hexagon;

    (xvi) a regular heptagon;

    (xvii) a non-regular heptagon;

(xviii) a regular octagon;

 (xix) a non-regular octagon;

 (xx) any *n*-sided figure for $n \geq 9$.

(If you haven't got wood and a saw, use polystyrofoam and a knife. Or just use your imagination. Or use http://www2.nzmaths.co.nz/frames/ brightsparks/tonisTiara.asp?applet. If you turn to the Bright Sparks section you will find a piece of animation that might speed up your work.)

Actually this whole section was inspired by a junk. I was cruising around Hong Kong harbour in a junk in mid-1990, with members of the New Zealand IMO team. We were on our way home from the 31st IMO in Beijing. On the rails around the junk there was a particular type of truncated cube.

For the moment then, concentrate on truncating a cube in the following way. Take a plane slice which cuts off one vertex or corner of the cube. Now do exactly the same to every other vertex of the cube.

Exercises

43. How can you form a truncated cube which has eight equilateral triangular faces and six regular octagonal faces?

 If the original cube has side length *t*, what is the length of one of the sides of the regular hexagon?

 How much of the original cube was cut away in this truncation?

44. What shape do the planes enclose, which produce the truncated cube of the last exercise?

 What is the volume of this shape?

45. Is it possible to truncate the cube so that it has eight equilateral triangular faces and six square faces?

 If the side length of the original cube was *t*, what is the side length of one of the square faces?

 How much of the original cube was cut away in this truncation?

46. Is it possible to truncate the cube so that eight regular hexagons are formed?

 What is the shape of the remaining six faces?

47. Is it possible to truncate the cube so that only an octahedron remains?

 What fraction of the volume of the original cube remains?

48. Investigate other truncations of the cube.

 Investigate truncations of the octahedron.

 Now try the dodecahedron.

 What about the other Platonic solids?

7.5. Chessboards

Chessboards are made up of squares and that is how I've managed to sneak them into this chapter. We've already had a crack at knight's tours in Chapter 3 of *First Step* and we've placed non-taking rooks on peculiarly shaped boards to good effect in Chapter 6. Now let's try another tack.

Exercises

49. What is the smallest number of rooks which can be placed on a chessboard so that every square is controlled by at least one of them?
    (As usual, if you can't do this first use a well tried problem solving technique. Try the smaller cases of $2 \times 2, 3 \times 3$ and 4×4 boards.)
    What is the answer on an $n \times n$ board?
50. What is the smallest number of bishops which can be placed on a chessboard so that every square is controlled by at least one of them?
    What is the answer on an $n \times n$ board?
51. Repeat the last exercise with bishops replaced by queens.
52. Repeat the last exercise with queens replaced by knights.
53. What is the greatest number of non-taking queens that can be placed on an $n \times n$ board?
54. What is the greatest number of non-taking bishops that can be placed on an $n \times n$ board?
55. Replace bishops by kings in the last exercise.
56. Replace kings by knights in the last exercise.

Moving on quickly, consider a 3×3 board. Place pawns in the dead centre of squares so that no three pawns are in a straight line — any straight line.

Exercises

57. How many pawns can be placed on a 3×3 board so that no three pawns are in a straight line?
    (The pawns are always placed at the centres of the squares.)
58. Repeat the last exercise for a 4×4 board.
59. Repeat the last Exercise for $5 \times 5, 6 \times 6, 7 \times 7, \ldots, n \times n$ boards.
60. In how many ways can the maximum number of pawns in each of Exercises 57–59 be placed?

This last set of exercises is about what is known as the ***no-three-in-a-line problem***. It's been around now since it was posed by Henry Dudeney in 1917 but nobody has yet come up with a complete solution. The obvious conjecture is that you can place $2n$ pawns on an $n \times n$ board so that no three are in any sort of line. By the Pigeonhole Principle (see Chapter 2 of *First Step*), the magic number can't be $2n + 1$. Currently the best lower bound estimate is a little under $3n/2$. A conjecture of Guy and Kelly for the upper bound has been corrected by Gabor Hellman to $\pi n/\sqrt{3}$. At the time of writing this then, no one knew the exact answer but they do know that it is between about $3n/2$ and $2n$.

Naturally if you can't do the no-three-in-a-line problem, extend or generalise.

Exercises

61. What progress can you make with the no-four-in-a-line problem?
62. How about a $3 \times 3 \times 3$ cube? Divide this up into 27 small cubes. Place as many fleas at the centres of the smaller cubes as you can (at most one flea per centre) so that no three fleas are in a straight line.
    Repeat for an $n \times n \times n$ cube.
63. How many fleas can you choose so that no three are in a plane?

Now you can think of dominoes as being 1×2 rectangles. It's clear from all that we did in Section 7.3 that you can cover an 8×8 chessboard with dominoes.

Exercises

64. Remove two diagonally opposite squares from a chessboard. Can you cover what remains with dominoes?
65. Which pairs of squares can you remove from a chessboard so that what remains cannot be covered by dominoes?
66. Take all the corner squares from a chessboard. Can you cover all the remaining squares with 1×4 rectangles?
67. Nine pieces forming a 3×3 square are placed in the lower left hand corner of an 8×8 chessboard. Any piece may jump over one standing next to it into a free square, i.e. may be reflected symmetrically with respect to a neighbour's centre. Jumps may be horizontal, vertical or diagonal. It is required to move the nine pieces to another corner of the chessboard

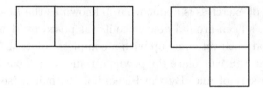

Figure 7.9.

(in another 3 × 3 square) by means of such jumps. Can the pieces be rearranged in the
(a) upper left hand corner?   (b) upper right hand corner?
(Tournament of the Towns, Junior Question No. 3, March 1987.)

Talking about dominoes, the 1 × 4 rectangle and squares, reminds me of polyominoes. These are dominoes with n squares. For instance, all the triominoes are shown in Figure 7.9.

Exercises

68. How many different tetrominoes are there?
69. How many different pentominoes are there?
70. How many different polyominoes are there?
71. Show that a 3 × 3 square can be made using two of one triomino and one of the other.
72. Can you make up a rectangle using all the different tetrominoes once and only once?
73. Repeat the last exercise using pentominoes.

For those of you that are hooked on polynominoes I recommend S. W. Golombs' book "Polyominoes", George Allen and Unwin, London, 1965. For the rest of you, read on.

Exercises

74. Now go into three dimensions. Call a *polyoid* a connected entity made up of n cubes so that any pair of adjacent cubes have a face in common.
    (i) How many dominoids (2 cubes) are there?
    (ii) How many trioids are there?
    (iii) How many tetroids are there?
    (iv) Generalise.
75. Can polyoids be joined together to form cuboids?

76. What's so fancy about squares? Let me call a ***polytriang*** a connected thing-a-me made up of *n* equilateral triangles where any pair of adjacent triangles has a whole edge in common.

    Explore the ideas from Exercises 68–73 using polytriangs.

I'm not really sure what polyominoes really have to do with chessboards. They just crept in while my back was turned. They do happen to be surprisingly simple, yet complicated animals. There is a lot we don't yet know about them. Maybe one of you will discover something new.

If you want to see some more questions related to chessboards there is a series of twenty-three articles published by the New Zealand Mathematics Magazine between 1991 and 2005 (written by Zulauf and Holton) that might interest you.

7.6. Squigonometry

In Chapter 7 of *First Step*, I defined the trigometric functions sine and cosine using a diagram something like the one below (see Figure 7.10).

So $\sin \theta = y$ and $\cos \theta = x$. We can also write $\tan \theta = y/x$ and find similar expressions for the various other trigonometric functions. But what is so special about circles? Why can't we define some new functions starting with squares? So let's invent squigonometry. From Figure 7.11, I'll define squine and cosquine.

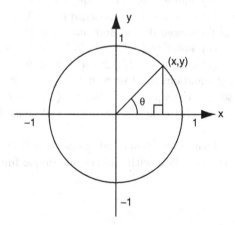

Figure 7.10.

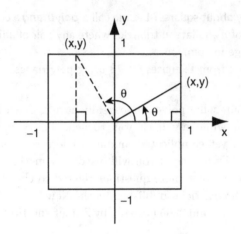

Figure 7.11.

Define squine θ = sqin θ = y and cosquine θ = cosq θ = x. What properties do these new functions possess?

Exercises

77. Are squine and cosquine bounded functions? In other words, are there some real numbers M, N, M', N' such that $M \leq$ sqin $\theta \leq N$ or such that $M' \leq$ cosq $\theta \leq N'$, for all θ.

78. Are squine and cosquine periodic? If so, what is the period?

79. Solve the following equations for θ between 0° and 360° inclusive.
    (i) sqin $\theta = 1$;   (ii) sqin $\theta = 0$;   (iii) sqin $\theta = -1$;
    (iv) cosq $\theta = 1$;   (v) cosq $\theta = 0$;   (vi) cosq $\theta = -1$.

80. For what values of θ between 0° and 360° inclusive, is
    (i) sqin $\theta \geq 0$;   (ii) sqin $\theta = 1/\sqrt{3}$;   (iii) sqin $\theta = -1/\sqrt{3}$;
    (iv) cosq $\theta < 0$;   (v) cosq $\theta = -1/\sqrt{3}$;   (vi) cosq $\theta = -1$?

81. Solve the following equations for θ such that $0° \leq \theta \leq 360°$.
    (i) sqin $\theta =$ cosq θ;   (ii) sqin $\theta = -$cosqθ;   (iii) sqin $\theta = \sin \theta$;
    (iv) cosq $\theta = \cos \theta$;   (v) cosq$^2\theta =$ cosq θ;   (vi) squin$^2\theta = -$sqin θ.

There are a whole host of questions that should now be flooding through your brains. Follow the parallels with the trigonometric functions and see where they lead you. Try to ask them before I do. In the meantime, let me define the tanquent function. Obviously

$$\text{tanquent } \theta = \text{tanq } \theta = y/x.$$

Exercises

82. Is tanquent a bounded function?
83. Is tanquent periodic? If so, what is its period?
84. Find θ such that $0° \leq \theta \leq 360°$ when
    (i) tanq $\theta = 0$;   (ii) tanq $\theta = 1$;   (iii) tanq $\theta = -1$.
85. For what values of θ is
    (i) tanq $\theta > 0$;   (ii) tanq $\theta > 1$;   (iii) tanq $\theta < -1$?
86. Solve the following equations for θ such that $0° \leq \theta \leq 360°$.
    (i) tanq $\theta = $ sqin θ;   (ii) tanq $\theta = $ cosq θ;
    (iii) tanq $\theta = \tan \theta$;   (iv) tanq$^2 \theta = -$tanq θ.
87. Sketch the graphs of sqin θ, cosq θ and tanq θ in the interval $0° \leq \theta \leq 360°$.
88. (a) Now $\sin^2 \theta + \cos^2 \theta = 1$. What is the corresponding relation for sqin θ and cosq θ?
    (b) What is the squigonometric equivalent of $1 + \tan^2 \theta = \sec^2 \theta$?
    (c) Is there a squigonometric double angle formula?
       Could sqin $2\theta = 2$ sqin θ cosq θ?
89. Those of you who have done some calculus could try to find the derivative of sqin θ with respect to θ. Is sqin θ differentiable for all θ?
    What about cosq θ and tanq θ?
90. The inverse relations for the trigonometric functions have a variety of notations. I'll use $\cos^{-1}$ and $\sin^{-1}$ to be the inverse **relations** such that $\cos^{-1}(\cos \theta) = \theta$ and $\sin^{-1}(\sin \theta) = \theta$.
    Sketch the graph of cosq^{-1} and sqin^{-1}, the inverse relations of cosq and sqin.

But, of course, I could have oriented the square of Figure 7.11 differently. Another symmetric way of using the square is given in Figure 7.12.

Define the sdine, cosdine and tandine functions from the diagram of Figure 7.12. (Here d stands for diamond. So if you like this is diamonometry.)

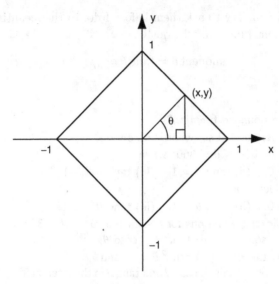

Figure 7.12.

Exercise

91. This is the basis for an interesting investigation. Here are some things that you might like to try.

    Repeat Exercises 77–90 using the diamonometric functions instead of the squigonometric functions.

    Are there any relations between squine and sdine or cosquine and cosdine? Why stick to squares? Put a nice rectangle with its centre at the origin and its axes parallel to the horizontal and vertical axes. Do some rectonometry. Try arbitrary closed shapes about the origin. Do you get any "nice" functions?

7.7. The Equations of Squares

In the section on Modulus in Chapter 7 of *First Step* we discovered that it was possible to find the equations of some squares. For instance, the graph of Figure 7.13(a) has equation $|x| + |y| = 4$ and that of Figure 7.13(b) has equation $|x - y| + |x + y| = 4$. So the question is, can we find the equation of any given square?

What I'd like to be able to do is to find the equation of any square that is centred at the origin. I figure I can find any other square by starting with a square whose centre is the origin and then transforming it. So what is the

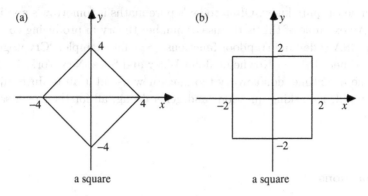

Figure 7.13.

equation of the square with vertices at $(1, 2), (2, -1), (-1, -2)$ and $(-2, 1)$? As with almost everything I do, I'll get to that eventually. Before I do get around to it, though, try these exercises and/or Exercises 19–25 of *First Step*, Chapter 7.

Exercises

92. Sketch the following graphs
    (a) $|x - y| + |x + y| = 2$;   (b) $|x - 2y| + |2x + y| = 5$;
    (c) $|x - 3y| + |3x + y| = 10$.
93. For the graphs of the last exercise, what is the significance of the $ax \pm by$ terms that are in the modulus signs?
94. Find the equation of the square with vertices at $(1, 2), (2, -1), (-1, -2)$ and $(-2, 1)$.
95. Show how to find the equation of the graph of any given square.
96. Show how to find the equation of the graph of any given parallelogram.

As usual you will find the theory behind this set of exercises in the solutions to them. But after you have gone through that work you should be able to find the graph of any equation of the type $|ax - by| + |ax + by| = c$ or find the equation of any square. Then, of course, I will expect you to be able to generalise that.

Anyway, I hope as a result of this chapter you will now see that mathematicians will never be out of business. Most answers suggest new problems. Most results can be generalised or extended in some way. True, a lot of these will be generalisations for generalisation's sake. They may have no immediate application or interest. The difficulty with mathematics though is that you never quite know. Often today's pure maths is tomorrow's applied maths. An example of this is the use of number theory in producing codes. Look up RSA codes and trapdoor functions. (See, for example, "Cryptography, A Primer" by A G Kowheim, John Wiley and Sons, New York, 1981.)

In the meantime, don't worry too much how useful it all is. Just enjoy solving problems, asking questions and generalising, all for their own sake and value.

7.8. Solutions

1. I first want to make the point that a square is a plane figure. Then I want to see if I can reduce the definition a little. Will two adjacent equal sides do? If all sides are equal how many right angles do you need to specify?

2. If you know anything about vectors you should do this problem easily. If not, you'll probably have to use Pythagoras' Theorem a bit.

   Now $UV^2 = 3^2 + 3^2 = 18 = XY^2$. But UV is parallel to ED and XY is parallel to FC. Since ED, FC are diagonals on opposite faces of the cube, they are parallel. Because UV and XY are equal and parallel, $XUVY$ is a parallelogram.

   Further, $EX^2 = EF^2 + FX^2 = 4^2 + 1^2 = 17$ and $UX^2 = 17 + 1^2 = 18$. So $XUVY$ is a rhombus.

   By Pythagoras again, you can show that $VX^2 = 36$ (go from V to X via U and F). Hence $XUVY$ has at least one angle of 90° since $VX^2 = VU^2 + UX^2$. So $XUVY$ is a square.

3. (a) 18; (b) $1/4pq(p+1)(q+1)$.

4. How can you pick a particular rectangle out of all of the possible rectangles in the $p \times q$ rectangle? All you need to do is to choose two pairs of opposite sides. So think of the $p \times q$ rectangle being made up of pq unit squares and this will break the rectangle up with horizontal and vertical lines. You can now get the opposite sides of any rectangle by choosing any two of the $p + 1$ horizontal lines and any two of the $q + 1$ vertical lines. The first pair of lines can be chosen in $^{p+1}C_2$ ways and the second pair in $^{q+1}C_2$ ways. So the total number of rectangles is $^{p+1}C_2 \, ^{q+1}C_2 = 1/4 - pq(p+1)(q+1)$.

5. (a) There are four ways of choosing four consecutive rows of the 7×9 rectangle. Within each of these there are five ways of choosing five consecutive columns. Hence there are twenty rectangles.

   (b) $(p + 1 - a)(q + 1 - b)$.

6. There are $(p+1-a)(q+1-b)$ rectangles of size $a \times b$ and $(p+1-b)$ $\times (q+1-b)$ rectangles of size $b \times a$. So we require $(p+1-a)(q+1-b)$ to equal $(p+1-b)(q+1-a)$. This reduces to $(b-a)[(p+1)-(q+1)] = 0$. (i) Hence $a = b$ or $p = q$. (ii) If the equation is true for all a and b, then p must equal q.

7. The "$1\times$" rectangles are $1 \times 1, 1 \times 2, \ldots, 1 \times q$ — there are q of these. The "$2\times$" rectangles are $2 \times 2, 2 \times 3, \ldots, 2 \times q$ — there are $q - 1$ of these. This continues until we get to the $p \times p, p \times (p+1), \ldots, p \times q$ rectangles — there are $(q + 1 - p)$ of these. Altogether the number of types is $q + (q-1) + (q-2) + \cdots + (q+1-p) = 1/2[q + (q+1-p)]p = 1/2p(2q+1-p)$. (Is there a quicker way of doing this?)

8. (a) 9; (b) 12.

   (c) Divide the equilateral triangle of sidelength n into equilateral triangles of sidelength 1. There are $2[1+2+\cdots+(n-1)]$ rhombuses of sidelength $1, 2[1+2+\cdots+(n-3)]$ of sidelength $2, 2[1+2+\cdots+(n-5)]$ of sidelength 3 and so on. The largest rhombus that can be fitted in has sidelength the integer part of $1/2n$.

   That should keep you busy getting a nice answer.

   Are trapezia (assuming opposite parallel sides are unequal) any better?

   How many parallelograms can you find? That's not quite as easy as the rectangles in a rectangle is it?

9. This is comparatively easy. It's just the natural three-dimensional extension of Exercise 4. The answer is ${}^5C_2{}^5C_2{}^5C_2 = 1000$.

10. If $n = 2$ do you get eight smaller tetrahedra? If you did, you need to check things again. You can actually fit one tetrahedron into each corner. Then you have a space in the shape of an octahedron. Unfortunately you can't carve that up into tetrahedra.

    There's a nice reference to this in Peter Hilton and Jean Pedersen's "Build your Own Polyhedra", Addison-Wesley, Menlo Park, 1988. Look for Jennifer's puzzle.

11. It's easy to see that there are 12 — one on each edge.

    But don't stop here. How many small cubes have red paint on 0, 1, 3, more faces?

    What happens if we start with an $n \times n \times n$ cube? What's so special about cubes?

12. Well it's 50 or 52 or 68 or 72 or 76. On the outer layer of one cube thick you can remove 6×5 cubes. This should leave you with a $3 \times 3 \times 3$ cube that ought to be easy to count.

    Experiment with other designs and an $n \times n \times n$ cube.

What is the smallest pattern that will remove **all** of the small squares?

13. Is the cube the only one that works?

   Any perfect cube will work for cubes.

14. What did you get?

15. (i), (ii) are the only ones. For the others, pq is odd. In these cases we can always make sure that only one square is left over. (Fit in as many vertical, or 2×1, rectangles as you can and then fill the remaining row with 1×2 rectangles to leave one square uncovered.)

16. **Conjecture.** *The 1×2 rectangles fit exactly if pq is even. There is one square uncovered if pq is odd.*

   Proof. If pq is even, then suppose without loss of generality that p is even. Then fit 2×1 rectangles in parallel to the even side, to completely cover the larger rectangle.

   If pq is odd then both p and q are odd. Fit in the rectangles as in the last exercise to leave a gap of one square at the end of every row. Then fill in this strip so that one squre of the rectangle is left uncovered.

17. (i) 1 square; (ii) 2 squares; (iii) no squares.

18. **Conjecture 1.** *If $pq \equiv \alpha \pmod 3$ then α squares will be left uncovered, where $\alpha \in \{0, 1, 2\}$.*

   Unfortunately this conjecture is false. Find a small counterexample.

   Conjecture 2. *If $pq \equiv \alpha \pmod 3$ then α squares will be left uncovered, where $\alpha \in \{0, 1, 2\}$, unless $p = q = 2$ when 4 squares will be left uncovered.*

   Proof. You can check directly for $1 \le p \le q \le 5$. Then, for $5 \le r \le s$, the wastage for the $r \times s$ rectangle isn't smaller than that for the $(r + 3) \times s$ rectangle. Keep going by Mathematical Induction.

19. Presumably a 6×6 rectangle can be completely covered by 1×4 rectangles since $36 \equiv 0 \pmod 4$. Why do I get 4 squares over then? Read the text for more. (Or try to do better with a 10×10 rectangle first.)

20. (i) 2; (ii) 4; (iii) 6 — but that's bigger than 5! (iv) 4; (v) 2; (vi) 6.

21. Things look alright except for (iii) and (iv). What happens in a 7×13 rectangle? Do we still have a wastage of 6?

   It turns out that if $p = 5k + a$ and $q = 5l + b$, where $k, l \ge 1$, then the wastage is

$$W = \begin{cases} ab, & \text{if } a + b \le 5, \\ (5 - a)(5 - b), & \text{if } a + b \ge 5. \end{cases}$$

Can you prove that?

(By the way, you need special arguments for p or q less than 5.)

What is the largest wastage possible? Is it really 16?

22. Let $p = kn + a$ and $q = ln + b$, where $k, l \geq 1$ and $0 \leq a, b < n$. Then the wastage W is given by

$$W = \begin{cases} ab, & \text{if } a + b \leq n, \\ (n-a)(n-b), & \text{if } a + b \geq n. \end{cases}$$

(A proof of this can be found in the paper by S Barnett and G J Kynch, "Solution of a simple cutting problem", Operations Research 15, 1967, 1051–1056. Your librarian might be able to track this down for you.)

23. (a) The answer here is quite complicated. Let n be odd, $p = kn + a$ and $q = ln + b$, where $k, l \geq 3$ and $0 \leq a, b < n$. let W' be the wastage and W be the wastage in the $1 \times n$ case.

Then

$$W' = \begin{cases} W, & \text{if } pq - W \text{ is even,} \\ W + n, & \text{if } pq - W \text{ is odd.} \end{cases}$$

If $1 \leq k, l \leq 2$ then W' can be W or $W + n$, but there seems to be no simple pattern here.

(This is proved in D A Holton and J A Rickard "Brick packing" in Combinatorial Mathematics, Lecture Notes in Mathematics, No. 686, Springer-Verlag, Berlin, 1978, 174–183.)

(b) Let $p = 3k + a$ and $q = 3l + b$, where $0 \leq a, b < 3$. Is the wastage really $pb + qa - ab$?

24. I'm afraid it can. What can you do with squares of side length 2 in a square of side length 5?

25. I honestly have no idea what the answer is to this one. I'm not even sure what wastage you get when an $m \times n$ rectangle is packed into a $p \times q$ rectangle.

26. (i) In this case the length of the diagonal AC is equal to T.

But $AC = \sqrt{(AB^2 + BC^2)} = t\sqrt{2}$. So $T = t\sqrt{2}$.

(ii) Here the length of the diagonal $A'C'$ is t. So $T = t/\sqrt{2}$.

27. (a) A little transformation geometry makes it obvious. Reflect about $A'C'$ and $B'D'$.

If you don't like that argument, let X be the point of intersection of OA' and AB. Since $\angle XA'P = 45°$ and $\angle A'XP = 90°$, then $\angle A'PX = 45°$. Hence $AX = XP$. Similarly $A'X = XQ'$. But

$AX = XB$, so $AP = AX - XP = XB - XQ' = Q'B$. $\triangle$'s APP', PQQ' are right angled triangles with two angles of size $45°$ (since $\angle AP'P = \angle PA'X$, alternate angles on parallel lines $AD, A'C'$). So $\triangle$'s APP', BQQ' are similar. But $AP = Q'B$. So $\triangle$'s APP', BQQ' are congruent. They therefore have equal areas.

(b) Straightforward by transformation geometry or via the argument of (a).

(c) If $\triangle APP'$ has the same area as $\triangle A'PQ'$, then $AP \cdot AP' = A'P \cdot A'Q'$. But both triangles are isosceles so then $AP = AP' = A'P = A'Q'$. Let this length be s. Then $t = AP + PQ' + QB = s + s\sqrt{2} + s$. Further $T = D'P' + P'P + PA' = s + s\sqrt{2} + s$. Hence $T = t$.

28. Let $AP = s$. Then the sum of the areas of the four triangles is $2s^2$. The area of the octagon is therefore both $2s^2$ and $t^2 - 2s^2$. So $t = 2s$.

     But in this case the octagon is a square!!

29. (i) In Figure (a) below, $OA' = T\sqrt{2}$ and $OA = t\sqrt{2}$.

     Further $\angle OAA' = 135° - \theta$. By the sine rule,

$$\left(\frac{1}{2}\right) T\sqrt{2} / \sin(135° - \theta) = \frac{\frac{1}{2}t\sqrt{2}}{\sin 45°}.$$

Since $\sin(135° - \theta) = (1/\sqrt{2})(\sin\theta + \cos\theta)$, $T = t(\sin\theta + \cos\theta)$.

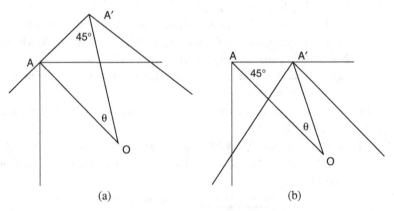

(a)   (b)

(ii) In Figure (b) above, $\angle OA'A = 135° - \theta$. Using the sine rule gives

$$(1/2)t\sqrt{2} / \sin(135° - \theta) = \frac{\frac{1}{2}T\sqrt{2}}{\sin 45°}.$$

Hence

$$T = \frac{t}{(\sin\theta + \cos\theta)}.$$

(Couldn't we have found an argument based on symmetry and (i) to get this?)

As A' moves down $A'O$ toward O starting in (a) and going to (b), the area of $\triangle A'P'Q'$ goes from positive to zero and the area of $\triangle APP'$ goes from zero to positive, continuously. Hence the difference in their areas goes from positive to negative, continuously. So somewhere in between the difference is zero. At this place the two areas are equal.

30. Transformation geometry is the easiest method.

31. Virtually the same argument as in Exercise 28 shows that this situation cannot happen.

32. (a) $T = t\sqrt{2}$; (b) $T = t/\sqrt{2}$.

Now the angle bisectors of an equilateral triangle are perpendicular to the opposite sides. But the perpendicular bisectors of the chords AB, AC, BC to the circumcircle meet at the circumcentre. Hence the incentre and the circumcentre are the same point.

By the symmetry of the equilateral triangle, the bisectors of $\angle$'s A, B, C are also the medians of $\triangle ABC$. Hence the incentre and the centroid are the same point.

By what I have already said, the angle bisectors of $\triangle ABC$ are also the altitudes of $\triangle ABC$. The altitudes meet at the orthocentre. So the orthocentre is the same point as the incentre.

33. Since AC, AB are tangents to the incircle, the incentre must lie on the bisector of $\angle A$. The same can be said for $\angle B$ and $\angle C$.

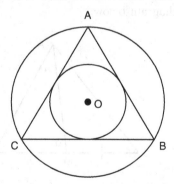

Now the angle bisectors of an equilateral triangle are perpendicular to the opposite sides. But the perpendicular bisectors of the chords AB, AC, BC to the circumcircle meet at the circumcentre. Hence the incentre and the circumcentre are the same point.

By the symmetry of the equilateral triangle, the bisectors of ∠s A, B, C are also the medians of $\triangle ABC$. Hence the incentre and the centroid are the same point.

By what I have already said, the altitudes meet at the orthocentre. So the orthocentre is the same point as the incentre.

34. (i) In the diagram, let T be the perpendicular bisector of AB.

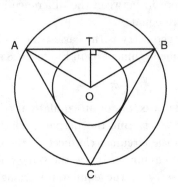

Suppose the incentre and the circumcentre are the same point, O. Then the perpendicular bisector of AB goes through O. Now $\triangle$'s OTA, OTB are congruent since $OT = OT, AT = TB$ and $AO = BO$ (radii of the circumcircle). Hence $\angle TAO = \angle TBO$.

But $\angle TAO = 1/2\angle A$ and $\angle TBO = 1/2\angle B$. Hence $\angle A = \angle B$. Similarly, $\angle B = \angle C$. So the triangle is equilateral.

(ii) Consider the diagram below.

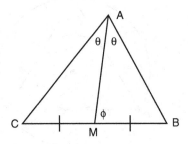

Suppose that the incentre and the centroid are the same. Then the bisector of $\angle A$ is the median AM. Using the sine rule on $\triangle$'s AMC, BMC we find that either $\theta = 180°$ (which is clearly impossible) or $\varphi = 90°$. Hence $\triangle ABC$ is again equilateral.

(iii) The bisector of $\angle A$ can only be an altitude if $\angle B = \angle C$ (see the diagram below).

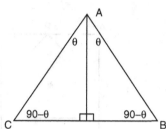

Similarly the bisector of $\angle B$ can only be an altitude if $\angle A = \angle C$. Hence $2\theta = 90° - \theta$. So $\theta = 30°$. This shows that if the incentre is the same point as the orthocentre, then the triangle is equilateral.

(iv) If the circumcentre and the centroid are the same, then the altitudes must be medians. This implies the triangle is equilateral.

(v) If the circumcentre and the orthocentre are the same, then the altitudes must again be medians.

(vi) So suppose the centroid and the orthocentre are the same. Again the altitudes are medians.

35. (a) The diagram is shown below. Clearly $T = 2t$.

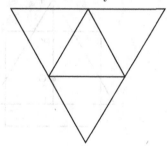

(b) In the "opposite" of the situation above we get $T = 1/2\,t$.

36. Surely $T = t$.

37. The area of an equilateral triangle with sidelength l is $\frac{1}{4}l^2\sqrt{3}$. Let Δ PA'B have sidelength l. Then $\frac{3}{4}l^2\sqrt{3} = \frac{1}{2}\frac{1}{4}t^2\sqrt{3}$. So $l = \frac{t}{\sqrt{6}}$. Hence $T = l + 2(t - 2l) = 2t - 3l = t(2 - \sqrt{(3/2)})$.

38. For the triangles to have equal area they have to be congruent. This leads to $T = t$.

(What can you say about the hexagon $AA'BB'CC'$?)

39. (a) $T = 2t$. (b) $T = 1/2\,t$.

40. You're on your own here.

41. (a) By the position of the centre of $\triangle PQR$, $OP = 2/3\,h$, where h is the height of the triangle.

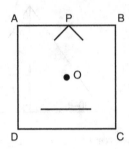

But $h = (e\sqrt{3})/2$ and $OP = 2/3\,h$, so $e = (s\sqrt{3})/2$.

   (b) Surely just $s = e$.

   (c) Use continuity arguments. You see we can arrange for $\triangle PL'M$ to have zero area and for $\triangle ALL'$ to have zero area.

   (d) Hmm. That should keep you out of mischief for a while.

   (e) Use the diagram below.

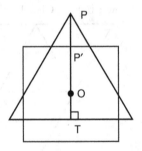

Now $OP' = 1/2\,s$ and $OT = \frac{1}{3}(\frac{e\sqrt{3}}{2})$. So $CN = 1/2\,s - OT$. On the other hand, $PP' = \frac{2}{3}(\frac{e\sqrt{3}}{2}) - \frac{1}{2}\,s$. Now plough on.

   (f) You must have worked out how to find the area of $\triangle ALL'$ in (d).

   (g) If (a) is true, then Q is inside the square.

   (h) Is $\frac{\sqrt{3}}{3} - \frac{1}{2}$ positive? If so, P is outside the square.

   (i) and (j) don't look too easy.

   (k) Well, who knows?

42. (i) to (vi) are fairly straightforward with only the right angled triangle not being possible. The doable triangles can be done by slicing off a vertex of the cube. Varying the angle gives the range of triangles.

A right angle can only be formed by taking a slice parallel to an edge of the cube. But such a slice can't produce a triangle;

(vii) One way to do this is to cut parallel to a face (but what other ways are there?);

(viii) One way to do this is to cut at a slight angle to the cut in (vii);

(ix) No;

(x) Shave off an edge but do so so that the cut is not parallel to an edge;

(xi) No;

(xii) and (xiii) see diagram. Small changes will turn a non-regular pentagon into a regular one;

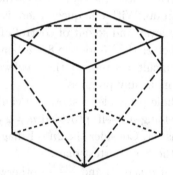

(xiv) and (xv) just move the last cuts a little so that the cut penetrates the sixth face;

(xvi), (xvii), (xviii), (xix), (xx) — is the hexagon as far as you can go? Has this got anything to do with the fact that a cube has six faces?

43. Intersect the cube with an octahedron. Suppose that the octagon has sides of length a. Then the sides that are not faces form a right angled isosceles triangle. So $t = a/\sqrt{2} + a + a/\sqrt{2}$. Hence $a = t/(\sqrt{2} + 1)$.

Now triangular pyramids are removed from each vertex. These have a base edge length of a and sloping edge of length $a/\sqrt{2}$. So the volume of one of these is $a^3/12\sqrt{2}$ to give a total volume removed of $a^3\sqrt{2}/3$.

44. See the last answer.

45. Make the equilateral triangles meet.

46. Keep cutting parallel to the cuts of the last exercise. The other faces should be squares.

47. Keep cutting parallel to the cuts of the last exercise until the squares become points. Then smaller and smaller octahedrons can be formed, all totally inside the cube.

48. You're on your own again.

49. You'll need n on the $n \times n$ board.

    (See A M Yaglom and I M Yaglom "Challenging Mathematical Problems with Elementary Solutions" Volume 1, Dover, New York, 1964 or look up the Rook Problem on the web. In fact the web is a good source of solutions for the exercises from here to Exercise 56.)

50. You'll need n again.

51. Ah. This is a bit more interesting. You only need two queens to cover a 4×4 board. (But these queens may be attacking each other.) This is the best you can do. With one queen there is always a square which is a knight's move away and so out of control of the queen.

    Can you do better than 6 for the 8×8 board?

52. This is very difficult because of the "awkward" moves that knights make. Did you make any progress?

53. Use the pigeonhole principle to show you can't get more than n. Then find a way of getting n itself, provided $n \neq 2$ or 3.

54. $2n - 2$ is the answer. Consider the $2n - 1$ diagonals on the $n \times n$ board, and especially the two shortest ones.

55. Here we get $\frac{n^2}{4}$ if n is even and $\frac{(n+1)^2}{4}$ otherwise.

56. It seems that there might be 32 on the standard chessboard!

57. Six is the most possible. One way is shown in the diagram below. How many others are there?

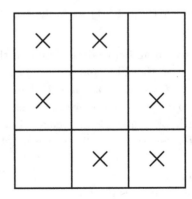

58. 8.

59. $2n$ might be a good conjecture. But the experts tend to think this is probably false. No one knows the precise answer. But a little more information is given in the text later.

60. Again, no one knows.

61. Probably you'll get $3n$, at least for small values of n. But who knows the general answer?

62. Well now, there's another fine mess I've got myself into. I really have no idea.

63. Ah. A trick question. Any three fleas lie in a plane, even if they are on a line.

64. No. A domino covers one white square and one black square. A lot of dominoes cover the same number of white squares as black squares. The butchered board of the question does not have an equal number of black and white squares.

65. A nice project. Clearly from the last exercise the squares will need to be of different colours. It's now a matter of going around all such pairs.

66. No. How can you cover one of the side squares next to a missing square? You'll either have to put the 1×4 rectangle horizontally or vertically. Follow the consequences of each such start.

67. First note that if a pawn starts on a white square it finishes on a white square. Similarly, a pawn on a black square always stays on black squares.

    (a) This is impossible. There are a different number of white squares in the lower left 3×3 square and the upper left 3×3 square.

    (b) Number the columns of the chessboard 1 to 8 starting from the left. Pawns which start on an odd column stay on an odd column. The same is true for pawns on an even column.

    Suppose the bottom left square is black. Then the number of black squares in the odd columns in the lower left 3×3 square is 4. The number of black squares in the odd column in the upper right 3×3 square is 1. The rearrangement is therefore impossible.

68. Four. They are shown below.

69. Twelve.

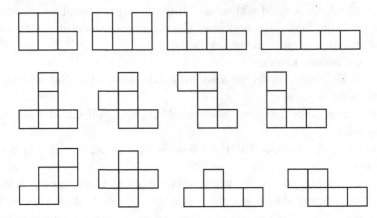

70. There are 35 distinct hexominoes and 108 distinct heptominoes. As far as I know, no one as yet knows the precise number of polyominoes with n squares. Counts go up to at least $n = 56$, where the number is just under 7 followed by 31 zeros.

71.

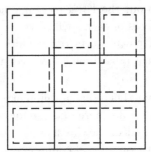

72. There are four tetrominoes so together they all cover 16 squares. The only rectangles with 16 squares have dimensions 1×16, 2×8 and 4×4. Trial and error shows that none of these can be covered in the desired way.

73.

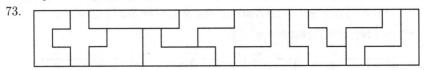

Above we show all 12 pentominoes forming a 3×20 rectangle. They will fit together to form another rectangle. What are its dimensions?

74. (i) 1; (ii) 2; (iii) 6 or maybe 7? (iv) I don't think anyone knows the answer to this.

75. Possibly not. But maybe so. What did you get?

76. Here are a few to whet your appetite.

77. Yes. The largest value of y is 1 and the smallest is -1.
Hence $-1 \leq$ sqin $\theta \leq 1$. Similarly $-1 \leq$ cosq $\theta \leq 1$.

78. After θ has increased by $360°$, the original values of sqin and cosq are repeated. This happens for no value smaller than $360°$. Hence the periods are $360°$.

79. (i) $45° \leq \theta \leq 135°$; (ii) $\theta = 0°$, $180°$; (iii) $225° \leq \theta \leq 315°$;
(iv) $-45° \leq \theta \leq 45°$; (v) $\theta = \pm90°$; (vi) $135° \leq \theta \leq 225°$.
(This is the first time we have seen any difference between the trigonometric and squigonometric functions.)

80. (i) $0 \leq \theta \leq 180°$; (ii) $\theta = 30°, 150°$; (iii) $\theta = 210°, 330°$;
(iv) $90° < \theta < 270°$; (v) $\theta = 120°, 240°$; (vi) $135° \leq \theta \leq 225°$.

81. (i) The x- and y-values are only equal at two opposite corners of the square. So $\theta = 45°, 225°$;

(ii) The other two opposite corners, so $\theta = 135°, 315°$;

(iii) $y = $ sqin $\theta = \sin\theta = \frac{y}{\sqrt{x^2+y^2}}$. So we get equality if $y = 0$ or $\sqrt{x^2 + y^2} = 1$. If $y = 0$, $\theta = 0°$, $180°$. If $x^2 + y^2 = 1$, then $y = 0$ and $x = \pm1$ or $y = \pm1$ and $x = 0$. In the latter case $\theta = 90°, 270°$. (This can be seen by comparing the defining circle for sin and the defining square for sqin.)

(iv) $x = $ cosq $\theta = \cos\theta = \frac{x}{\sqrt{(x^2+y^2)}}$ So $\theta = 0°, 90°, 180°, 270°, 360°$;

(v) cosq θ (cosq $\theta - 1) = 0$, so $\theta = 0°, 90°, 180°, 270°, 360°$;

(vi) sqin θ (sqin $\theta + 1) = 0$, so sqin $\theta = 0$ or -1. Hence $\theta = 0°, 180°, 270°, 360°$.

82. On the "top half" of the square $y = 1$ and x varies from 1 to -1. As $x \to 0, \frac{y}{x} \to \infty$. So tanq θ is not bounded. It can take any real value (positive or negative).

83. The period is $180°$.

84. (i) $\theta = 0°, 180°, 360°$; (ii) $\theta = 45°, 225°$; (iii) $\theta = 135°, 315°$.

85. (i) $0 < \theta < 90°$ and $180° < \theta < 270°$;
(ii) $45° < \theta < 90°$ and $225° < \theta < 270°$;
(iii) $135° < \theta < 180°$ and $225° < \theta < 360°$.

86. (i) $\frac{y}{x} = \tan q\,\theta = \text{sqin}\,\theta = y$. So $y = 0$ or $x = 1$. This gives $\theta = 0°$, $180°, 360°$ or $-45° \le \theta \le 45°$.

   (ii) $\theta = 45°, 135°$;

   (iii) Are they really equal for all θ?

   (iv) $\tan q\,\theta = 0$ or -1, so $\theta = 0°, 180°, 360°$ or $135°, 315°$.

87. (i)

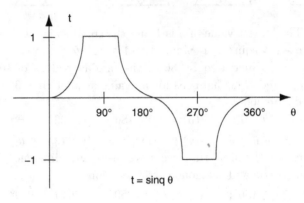

$$t = \text{sinq}\,\theta$$

(ii)

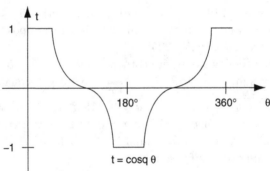

$$t = \text{cosq}\,\theta$$

(iii)

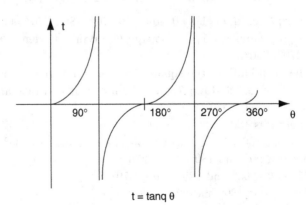

$$t = \text{tanq}\,\theta$$

88. (a) What is the equation of the square in Figure 7.10? (See *First Step*, Chapter 7, Exercise 19(ii) or wait until you get to section 7.7 of this chapter.)

   (b) $|1 - \tanq \theta| + |1 + \tanq \theta| = 2/|\cosq \theta|$;

   (c) Well $\sqin 2\theta \neq 2\sqin \theta \cosq \theta$ that's for sure.

89. No. Look at the graph at $\theta = 45°$. cosq and tanq are not differentiable either.

90.

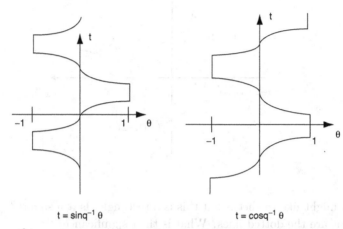

$t = \sinq^{-1} \theta$ $t = \cosq^{-1} \theta$

91. I leave this up to you and your imagination.

92. (a) To do this carefully, we need to consider the cases (i) $x - y \geq 0$ and $x + y \geq 0$; (ii) $x - y \geq 0$ and $x + y \leq 0$; (iii) $x - y \leq 0$ and $x + y \geq 0$; (iv) $x - y \leq 0$ and $x + y \leq 0$. From (i) we get $x = 1$; (ii) $y = -1$; (iii) $y = 1$; and (iv) $x = -1$. This gives the graph below, bearing in mind that each of the four equations we have just obtained are only valid for certain parts of the plane that depend on the inequality conditions.

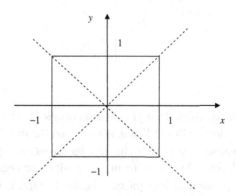

What are the equations of the dotted lines?

Now you can do Exercise 88(a).

(b) Again split the plane up into four regions depending on where the various absolute values are positive or negative. Then you should get the graph below.

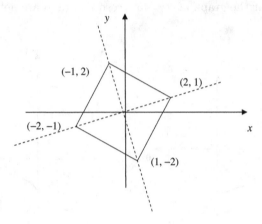

You might like to check that this is a rectangle. Is it a square? What are the dotted lines? What is their significance?

(c)

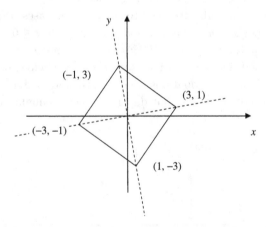

You might like to check that this is a rectangle. Is it a square? What are the dotted lines? What is their significance?

93. You should realise by now that the $ax \pm by$ terms are significant because it is by considering when these terms are positive or negative that we can break up the plane into four parts. The final graph is then a collection

of four lines that sit in these parts. In sitting there the graphs form a square with the lines $ax \pm by = 0$ as diagonals.

94. This time we start from the other direction. If we are to have a square it will have diagonals. These diagonals will break up the plane. So first find the diagonals. The line through (1, 2) and $(-1, -2)$ is $y = 2x$ and the line through (2, −1) and (−2, 1) is $2y = -x$. This means that the graph has equation $|2x - y| + |x + 2y| = c$. Here $c \geq 0$ because it is the sum of two moduli. But there are an infinite number of graphs like this. We now have to sort out the one we want. Now the graph passes through (1, 2) so substituting in the last equation gives us $|2 - 2| + |1 + 5| = c$. Since c has to be positive, c must be 5. The equation we want is

$$|2x - y| + |x + 2y| = 5.$$

You might want to check this out by starting again as we did in Exercise 92.

95. Using the last exercise as a model you can now find the equation of any square. I know that we have only looked at squares centred around the origin but the method of the last exercise will work no matter where the square is.

96. The method is exactly the same for parallelograms. (Put $c = 10$ in Exercise 94 and see what happens.)

The question now is can you find the equation of *any* quadrilateral this way? How about a regular hexagon? Or any hexagon? Or any polygon?

88. (revisited). After the last exercises you should be able to see the relation between sqin θ and cosq θ.

Chapter 8

IMO Problems 2

8.1. Introduction

This is the second look at some IMO problems. The first was in Chapter 8 of *First Step*, where I looked at some problems that were aired for the 1989 Olympiad in Braunschweig. The current problems start with some that were among those offered for the 1990 Olympiad in Beijing and finishes with a spray of some more recent ones that I leave you to tackle unaided, except for an odd hint and a commentary on the solutions that I hope might help. One of the aims of this chapter is to show that you can sometimes pull yourself up by the bootstraps. With a little (a lot of?) worrying, even IMO problems can fall to hard work and a bit of knowledge. Starting from nothing you can sometimes find a solution by sheer hard work. But you have to know how and where to look! Another aim is to show the scope and difficulty of IMO problems. If you aspire to that level of competition you will probably have to read widely and do a lot of work. But if you just enjoy solving problems and finding challenges, then this chapter is for you too.

Hopefully by looking at the sort of approaches I take here, the sort of approaches discussed in Chapter 3, you'll be able to see how you might tackle future problems you meet. The method is nearly always to try a few examples, experiment, see what patterns you can find, and hopefully in the process you will see both the answer to the problem and the proof that this is indeed the answer. Whatever happens you will have understood more of what mathematics is about and how infuriating it can be at times and how stimulating at others.

Now I know that in mathematics finding the answer is the fun part. Finding a proof is often hard work, even drudgery. However, until you've completed the proof you can't be sure that your answer is the correct and only answer. Unfortunately proof is an important part of mathematics. It's a part you have to get used to if you are going to make any real progress in the subject. And proof is emphasised again in this chapter.

Happy hard work.

8.2. AUS 3

At an IMO, the Jury consists of the team leaders of each country. It is the Jury that decides which six problems are used in the competition. They choose the six from a set, usually of about 30 problems, that are selected for them by the host country. In turn these 30 have been selected from all the problems submitted by all the countries.

To help identify the problems during the Jury's deliberations, each problem is labelled by a four character string. The first three characters are letters. These represent the country which submitted the problem. The final character is a number. This is the number of the problem submitted by the given country. So AUS 3 is the third question submitted by Australia. AUS 3 is listed as Problem 1 here.

Problem 1. *The integer* 9 *can be written as a sum of two consecutive integers:* $9 = 4 + 5$*; moreover it can be written as a sum of (more than one) consecutive positive integers in exactly two ways, namely* $9 = 4 + 5 = 2 + 3 + 4$*. Is there an integer which can be written as a sum of* 1990 *consecutive integers and which can be written as a sum of (more than one) consecutive positive integers in exactly* 1990 *ways?*

The idea now is that you try to solve Problem 1. When you've finished or when you're stuck, turn to Section 8.8, Hints AUS 3. This gives a series of problems that are related to AUS 3 and may shed some light on it. The complete solution, surprisingly, is in the Solution section, starting on page 272.

8.3. HEL 2

The problem HEL 2 was the second Greek problem.

Problem 2. *Let* $f(0) = f(1) = 0$ *and*
$$f(n + 2) = 4^{n+2} f(n + 1) - 16^{n+1} f(n) + n2^{n^2}, n = 0, 1, 2, \ldots$$
Show that the numbers $f(1989)$*,* $f(1990)$*,* $f(1991)$ *are divisible by* 3.

This is a recurrence relation problem. You might like to refresh you memory about such things by looking back at Chapter 1. In view of this, you might like to write $f(n)$ as a_n.

Unfortunately though there is nothing at all like this in Chapter 1. Can we invent some problems which are similar to Problem 2 that might give us a clue to solving this problem? When you've tried and failed, you might like to have a look at the Hints on page 263. When you've tried and succeeded, you might like to have a look at the Solution on page 279. Bear in mind though, that this is not the only way to solve this problem. If you have another method that you can justify, then I'd like to hear about it.

8.4. TUR 4

This is a nice problem. It was included as question 4 at the 31st IMO. In the Fragrant Hill Hotel on the outskirts of Beijing, I spent a lot of time on one of our students' solutions to this problem. At one stage I convinced myself that we had a perfect solution — maybe even a solution that was worth a special prize. The next morning, however, one of the other team leaders showed me the error of my ways. Well, the solution I had may not have been a complete solution but at least it was a step along the way. So what is the problem?

Problem 3. *Let Q^+ be the set of positive rational numbers. Construct a function $f : Q^+ \to Q^+$ such that*

$$f(xf(y)) = \frac{f(x)}{y} \quad \text{for all } x, y \in Q^+.$$

Of course TUR 4 was submitted by Turkey.

8.5. ROM 4

This is a nice little question from Romania which is a variant on a theme that we played as long ago as Chapter..., but that would be giving too much away.

Problem 4. *Ten localities are served by two international airlines such that there exists a direct service (without stops) between any two of the localities and all airline schedules are both ways.*

Prove that at least one of the airlines can offer two disjoint round trips each containing an odd number of landings.

I'm sure you can get that one out for yourselves. It just takes a little bit of persistence. If you're stuck though, go to the Hints starting at page 266. As usual I don't guarantee to get you home by the shortest route.

8.6. USS 1

This is a problem for those of you who like playing around with congruences or divisors. It's a pity though that there isn't a slightly more interesting answer.

Problem 5. *Consider the n-digit numbers consisting of 1 seven and $n - 1$ ones. For what values of n are all these numbers prime?*

I should point out, just in case you haven't guessed, that USS stands for the USSR. (USSR stands for the Union of Soviet Socialist Republics. This existed from 1922 until 1991 when it was divided into 15 countries.)

8.7. Revue

In this chapter I've tried to give you some idea of how to tackle some IMO-type problems. The ones I have chosen can be vanquished by a frontal attack plus a little insight. Actually most of life's problems can be solved this way, though often a little thought can find better solutions. But not surprisingly, knowledge is important. The more you know the easier it is to solve the next problem.

I've also tried to provide some revision of some of the ideas in earlier chapters. This is to make the point that nothing is ever done or learned in isolation. You can usually find a use for almost anything sometime. So arrange your mental filing system for ready access and retrieval.

You may have noticed that I don't have any geometry questions in the first part this chapter. That's not because I don't like, or can't do, geometry. (Although I must say that I feel happier with number theoretical or combinatorial problems.) Rather it's because I was looking for problems that could be developed and generalised. Perhaps because of my own interests or genes or whatever, or perhaps because it is intrinsically the case, I find it harder to develop and generalise geometry questions. Anyone who has this facility should contact me. I'd be happy to produce a chapter like this solely on geometry. (But the geometry is coming — in Section 8.13.)

Finally I've tried to give you some problems that, for some reason or other I've found interesting. Now I've really no idea what makes a problem interesting. It's undoubtedly subjective. It's just a matter of taste. I find it very hard to know why some topics, areas, problems somehow seem to have more appeal than others. Certainly, problems which are easily stated seem to have appeal. I'm sure that is why the Four Colour problem has always attracted a wide range of people, many of whom have no other mathematical interests. For me, a nice or surprising answer or an elegant twist, or a novel idea in the solution always helps. That's why I was a bit disappointed that Problem 5 came out the way it did. It would have held more fascination if, for say, $n = 5{,}223{,}119$ suddenly all the numbers turned up prime, right out of the blue.

Anyway, I hope you get as much enjoyment pouring over these problems as I did pouring them out.

8.8. Hints — AUS 3

This problem should ring bells for some of you. Any of you who have read *First Step* will have met the idea in Chapter 1 there. Let's recap. Suppose n can be written as the sum of k consecutive positive integers starting from r. Then

$$n = r + (r + 1) + (r + 2) + \cdots + (r + k - 1).$$

So n is the sum of an arithmetic progression with first term r and common difference one. Hence

$$n = 1/2\, k(2r + k - 1)$$

So

$$2n = k(2r + k - 1). \tag{1}$$

Note that in Problem 1 we were at pains to say "more than one". Obviously any number can be written as the sum of one consecutive number — itself. So we are only interested in $k > 1$. Actually we're eventually only going to be interested in $k = 1990$.

Exercises

1. Find k if
   (i) $n = 29$;   (ii) $n = 24$;   (iii) $n = 16$;   (iv) $n = 72$.
2. In Exercise 1, in how many sums of consecutive integers can each n be written?
3. If n cannot be written as a sum of (more than one) consecutive positive integer, what can be said about n?
4. If n can be written as the sum of consecutive positive integers in only one way, what can be said about n?
5. If n can only be written as the sum of consecutive integers in just two ways, what can be said about n?

6. Guess in how many ways $3^4 5^7$ can be written as the sum of consecutive positive integers.
   Prove that your guess is correct.

Rather surprisingly it's beginning to look as if the prime decomposition of n has got something to do with the number of ways it can be written as a sum of consecutive positive integers. The prime decomposition seems to come in because of equation (1).

Let's concentrate on n odd for a moment. From equation (1) we see that $2n = k(2r + k - 1)$. We know that k is the smaller of the two factors k and $2r + k - 1$. Further we know that precisely one of k, $2r + k - 1$ is even. So if we take any pair of factors s, t with $1 < s < t$ and such that $2n = st$, then $k = s$ gives a solution, provided one of s and t is odd and the other even. But if n is odd, precisely one of s and t must be odd too. The only time we don't get a solution is when $k = 1$. We must remember to throw this away at the end.

Exercises

7. (a) How many factors have the following numbers:
       (i) $n = 3.7^2$;   (ii) $n = 3^2 7^3$;   (iii) $n = 3^8 7^5$?
   (b) How many values are there for s in (a) if $n = st$ and $s < t$?
8. How many ways are there of writing the numbers n of the last exercise so that n is the sum of consecutive positive integers?
9. If $n = p^\alpha q^\beta r^\gamma$ where p, q, r are odd primes and $p < q < r$, how many ways are there of writing n as the sum of consecutive positive integers?
10. Repeat the last exercise with $n = 2^\alpha q^\beta r^\gamma$, where q, r are distinct odd primes.
11. Repeat Exercise 9 for $n = p_1^{\alpha_1} p_2^{\alpha_2} \ldots p_r^{\alpha_r}$ where $p_1 < p_2 < \cdots < p_r$ are distinct primes.

Let's return to Problem 1. Remember it was about slightly more than the number of ways n could be expressed as a sum.

Exercises

12. Is there an integer which can be written as a sum of two consecutive positive integers and which can be written as such a sum in exactly two ways? How many integers are there of this type?
13. Replace "two" by "four" in the last exercise.
14. Solve Problem 1.

15. You will note that IMO problem setters like to have problems that contain numbers that are the same as the year. Replace 1990 by the current year in Problem 1. What happens if you replace 1990 by any $m > 1$? How many solutions did you get? Is there any value of m that gives **no** solution?

8.9. Hints — HEL 2

There are two difficulties with the recurrence relation

$$f(n + 2) = 4^{n+2} f(n + 1) - 16^{n+1} f(n) + n 2^{n^2}. \tag{2}$$

The first difficulty is the 4^{n+2} and the 16^{n+1}. The second difficulty is the $n 2^{n^2}$.

Exercises

16. Solve $(n + 1)a_{n+1} = n a_n + (n - 1)a_{n-1}$, where $a_0 = a_1 = 1$.
17. Solve $2^{n+1}a_{n+1} = 2^n a_n + 2^{n-1}a_{n-1}$, where $a_0 = a_1 = 1$.
18. Solve $2^{n+1}f(n + 1) = 2^n f(n) - 2^{n-1}f(n - 1)$, where $f(0) = f(1) = 1$.

So those three exercises should give you some clue as to what to do with the 4^{n+2} and the 16^{n+1} in Problem 2. How far can you go with the problem now?

Exercises

19. How far can you go with Problem 2 now?
20. Try $f(n) = 2^{h(n)}g(n)$ in $f(n + 2) = 4^{n+2}f(n + 1) - 16^{n+1}f(n)$.
21. Now try Exercise 19 again.

In Chapter 4 of *First Step* (page 130) we came across a little trick that has been of some use to us a couple of times. If you add up a lot of equations that all have the same form, and if there are enough negative signs, and you are a Sagittarian (I'm an Aquarian myself), then lots of things cancel and simplification occurs.

Exercises

22. Consider the recurrence relation $g(n + 2) = 2g(n + 1) - g(n)$, where $g(0) = g(1) = 0$. Add the following equations and so find an expression for $g(n)$.

$$g(n + 2) = 2g(n + 1) - g(n),$$

$$g(n+1) = 2g(n) - g(n-1),$$

$$\cdots$$

$$g(4) = 2g(3) - g(2),$$
$$g(3) = 2g(2) - g(1),$$
$$g(2) = 2g(1) - g(0).$$

23. Solve $g(n+2) = 2g(n+1) - g(n) + 1$, where $g(0) = 0 = g(1)$.
24. Solve $g(n+2) = 2g(n+1) - g(n) + n + 2$, where $g(0) = 0 = g(1)$.
25. Solve $g(n+2) = 2g(n+1) - g(n) + (n+2)^2$, where $g(0) = 0 = g(1)$.
26. Have another look at Problem 2.

Suppose r is some constant. What is

$$\sum_{t=0}^{n} r^t = 1 + r + r^2 + \cdots + r^n?$$

This might take you back to Chapter 6 of *First Step* (page 189). It was here we first came across geometric progressions (G.P's).

So $\sum_{t=0}^{n} r^t = \frac{r^{n+1}-1}{r-1}$.

Exercises

27. Find $\sum_{t=0}^{n} 16^{-t}$.
28. Use calculus and the sum of a G.P. to find an expression for $\sum_{t=0}^{n} tr^{-t-1}$.
31. Simplify $\sum_{t=0}^{n} t16^{-t-1}$.
32. Find an expression for $2 + 6r + 12r^2 + \cdots + n(n-1)r^{n-2}$.
33. Go back to Problem 2 again.

Sorry about that. There are a couple of ideas in there that are worth remembering. Mostly it's a lot of slog, however. So let's try something a little nicer.

8.10. Hints — TUR 4

I don't know about these functional equation problems. I think the first thing that you need to do is to play around with them and try to get some feel for what's happening. Often it's worth trying to find $f(0)$ but of course that's no help here since the domain of f is Q^+.

Exercise

34. (a) Find $f(1)$, $f(2)$, $f(3)$.
    (b) Look for some relationship. Could $f(x) = x$ or $\frac{1}{x}$ or something else equally "simple"?

Well out of that experimenting we didn't get very far. In fact, the only ray of sunshine is the possibility that $f(1)$ may equal 1. But that requires f to be an injective function. That is, a 1:1 function, a function f such that two different values of x always give two different values of $f(x)$. Hence if $f(x_1) = f(x_2)$, then $x_1 = x_2$. Such functions are called *injections*.

Exercises

35. Which of the following functions $f: \mathbb{R} \to \mathbb{R}$, are injections?
    (i) $f(x) = x$;   (ii) $f(x) = x^2$;   (iii) $f(x) = x^3$.
36. Give two more examples of 1:1 functions.
37. Go back to Problem 3. Is the function in the problem a 1:1 function?

So let's recap now. We know that we have an injective function. We know that $f(1) = 1$. We know that $f(x) \neq x$ and $f(x) \neq \frac{1}{x}$. Where can we go from here?

Exercise

38. Let $x = 1$, $y = 2$. What is $f(f(2))$?
    Is that related to $f(f(1/2))$?
    Does this relation work for values other than 2 and $1/2$?

Let's try to absorb that. What we've shown is that repeated applications of f cycle values around. The image of y is $f(y)$. The image of $f(y)$ is $1/y$ The image of $1/y$ is $f(1/y)$. The image of $f(1/y)$ is y again. (Unless, of course, $y = 1$, in which case 1 goes to 1 goes to 1 goes to 1. Or rather, 1 just stays fixed where it is.)

Exercises

39. Find a candidate for $f(2)$.
40. If $f(2) = a$, what is $f(a)$? What is $f(f(a))$?
41. Suppose $b \neq 1$. If $f(b) = c$, what is the cycle of function values starting with b?
42. How arbitrary is all this. For any $b_1 \neq 1$, can we choose c_1 such that $f(b_1) = c_1$, provided $f(c_1) = 1/b_1$ and $f(1/c_1) = 1/b_1$. Then for any $b_2 \neq 1$ so far not used, can we choose c_2 such that $f(b_2) = c_2$, $f(c_2) = 1/b_2$, $f(1/b_2) = 1/c_2$, $f(1/c_2) = b_2$? And can we repeat this for $b_3, b_4, \ldots$?

I had a student who reached that stage of Exercise 42 and then stopped. For quite some time he thought that the function f defined in Problem 3 was so lax as to allow almost arbitrary choices. Alas, it isn't so.

Exercise

43. In the solution to Exercise 42 is it true that $f(2 \times 5) = f(2) \times f(5)$?
    Is $f(ab) = f(a)f(b)$ for **all** $a, b \in Q^+$?

Clearly this latest revelation reduces the possibilities for arbitrariness. But we still haven't found a function f that works. Is there one? Is there more than one?

Exercises

44. Before we find one function f, let us see how restrictive being a multiplicative function is.
    Let $n = p_1^{\alpha_1} p_2^{\alpha_2} \dots p_r^{\alpha_r}$, where p_i are prime numbers and α_i are non-negative integers.
    What does $f(n)$ rely on?
45. Let $m/n \in Q^+$, where m, n are positive integers. Is $f(m/n)$ defined uniquely when f is defined on the set of prime numbers?
46. Does this give you enough clues to find one f which satisfies the functional equation of Problem 3?

Fine so we've solved Problem 3. But is the function you found in Exercise 46 unique? If not, what do all solutions to Problem 3 look like?

Exercises

47. Find another solution to the functional equation of Problem 3.
    Find an infinite set of solutions.
48. How arbitrary are the images of the prime numbers in the functional equation of Problem 3?
    Does the image of a prime number **have** to be another prime number or the reciprocal of a prime number?
49. Can we find **all** functions which satisfy the functional equation of Problem 3?
50. Construct a function $f = Q^+ \rightarrow Q^+$ such that $f(xf(y)) = yf(x)$.
    Is f unique?

8.11. Hints — ROM 4

Surely, if you've learnt anything from *First Step* and this book, you just **have** to see that this is a graph theory problem. This problem is surely about colouring the edges of K_{10} either red or blue (one colour per airline) in an attempt to show that there must be two vertex-disjoint odd cycles in one colour only.

This sort of problem is in an area of graph theory called Ramsey Theory. This is because the first major result was proved by Frank Ramsey (http://en.wikipedia.org/wiki/Frank_P._Ramsey) who was surprisingly not a graph theorist. Another thing about him is that his younger brother became an Archbishop of Canterbury. A sad thing about him is that he died so young. But his theorem in combinatorics says that there is a smallest complete graph such that if you colour its edges in two colours it will contain either a complete graph of a given size in one colour or the other. You might try to find out why he was interested in such things. You might also try to find out why Erdös and Szekeres, who were graph theorists, proved the same theorem but didn't get it named after them. (Try looking at http://en.wikipedia.org/wiki/Erd%C5%91s%E2%80%93Szekeres_theorem. And then follow any leads from there. And what's the happy ending theorem?)

Well, having said that, there's not much more to prove!

Exercises

51. You should have done enough Ramsey Theory to know that K_{10} contains at least one odd cycle. How big is it?
52. Does that mean that there are two red triangles?
53. What happens if we are landed with a red triangle and a blue triangle? Can we get a monochromatic triangle in what's left?

That leaves us at something of an impasse. If we find ourselves with two disjoint triangles, one red the other blue, there's no room left to necessarily find an odd cycle in either red or blue.

Exercises

54. Suppose, for a moment, that we could wangle another free vertex somehow. If we had a K_5 could we force a monochromatic triangle on K_5? Could we force a monochromatic 5-cycle? If we could, that would be as useful for the problem. After all the problem only requires two disjoint odd cycles.
55. How can we squeeze a red triangle and a blue triangle into the first five vertices so that we can use the monochromatic triangle or 5-cycle of the lemma obtained from the last exercise?

You ought to know better than to think that having solved Problem 4 I'm going to be satisfied. You should by now know what questions I'm going to ask next. What are the natural questions to ask? What is the question on all your lips?

Exercises

56. Is 10 best possible in the context of Problem 4? In other words, are there smaller complete graphs which, when the edges are arbitrarily coloured red or blue, produce two disjoint triangles?

    What do you think of this argument?

    In Exercise 13 Chapter 2 of *First Step*, we showed that if we colour the edges of K_6 in two colours we either get two red triangles, a red and a blue triangle, or two blue triangles.

    Since K_9 has a triangle of some sort call it blue. Remove this triangle. In the K_6 left, we can use that Exercise 13 to get two disjoint triangles of one colour or the other. Hence we produce a monochromatic union of odd cycles.

57. What is the best possible number in the context of Problem 4?

58. What is the smallest n for which, when the edges of K_n are coloured red and blue, there is a monochromatic disjoint union of even cycles?

59. What is the smallest n for which, when the edges of K_n are coloured red, white and blue, there is a monochromatic disjoint union of odd cycles?

That should leave you enough to chew over. When you've settled all those questions go on to Problem 5.

8.12. Hints — USS 1

In this sort of problem it is usually worth playing around a bit first. It helps you to get a feel for the question at least.

Exercises

60. Have a look at the cases $n = 1, 2, 3$. For which of these values of n are all the proposed numbers prime.

61. How about $n = 4, 5, 6$?

62. What patterns have started to emerge?

    What conjectures do you have?

63. Try to prove some of your conjectures.

Most problems fall apart when you look at them in the right way. What's the key to this problem?

Exercises

64. Is there a convenient way to write every N formed from $n - 1$ ones and 1 seven?
65. We know that if $n \equiv 0 \pmod 3$ then none of the corresponding N are prime.

    For what n are there some N divisible by 2, 3, 4, 5, 6, 7, 8, 9, 10, 11?
66. Find N which are divisible by 7 for $n = 8, 10, 11, 13$ and 14.
67. Now solve Problem 5.

I think you would be surprised if I stopped there. As you all know by now a mathematician's work is never done. What questions could we ask now?

Exercises

68. What questions could we now ask?
69. Solve Problem A posed in the solution of Exercise 68.
70. Solve Problem F.
71. Solve Problems B, C, D, E, G.
72. Solve any of the questions that you thought of that I didn't.
73. Do you get anything interesting if you change the numbers in any of these questions to the current year?

8.13. Some More Olympiad Problems

In this section I want to give you access to some recent IMO problems. I have chosen the whole of the IMO problems for 2008 (the Madrid Olympiad), and a geometry question from IMO 2010. It is quite likely that the solutions that I present here will contain ideas that have not been covered in this book or in *First Step*. After all, the IMO is at the peak of school level mathematics competitions and the problems there are very hard and often very deep (and I haven't been able to cover all of the hard and deep material that it would be nice to know). But they will give you a chance to try your wings and see what you are capable of. If you see no obvious way to proceed think of what I said in Chapter 3. You can experiment even with geometry questions. If a problem involves a triangle, for instance, it might be worth looking at a right angled triangle or an equilateral triangle or some other triangle that might be easier to make progress with than the general one that is in the question. Working with specific triangles may give you the clue that you need for the general case.

If you can't seem to get a solution, then talk with a friend about it or ask your teacher for a hint. Only then look at the solution, and only at the start of the solution. What does the first sentence say? Does that give you a clue on how you might proceed?

To the solutions I have added a commentary. This is designed to try to make the reading easier by saying why a particular step was taken and giving some idea of the key points of the proof. I hope that this commentary will help you to read and understand the proof better. Reading maths is not easy. It needs practice. It is remiss of me to not have talked about reading maths solutions and proofs earlier. But I find it useful to read through a proof completely on the first read but with no great hope that I will understand everything. The aim of this first read is to get a feel for what the major steps are and where I am likely to have difficulties. I note these difficulties and come back on the second read with a view to looking at them in depth and to trying to understand them completely. All the while I have pen and paper handy. This may be to write down the steps in my own words or to try examples to see if the general statements make sense in a particular case. I also have books nearby, or access to the web available, so that I can look up anything that seems to refer to a proof or known fact that I did not know. If I am still stuck this may be due to an error in the proof (that doesn't happen often) or a misreading on my part. At this point it is often a good strategy to talk to a colleague to see if they can shed any light on the situation.

Before ending this discussion I should give you two URLs. These are http://www.imo-official.org/ and http://www.artofproblemsolving.com/ Wiki/index.php/IMO_Problems_and_Solutions,_with_authors. These are both sites that contain information about the IMO and IMO problems and solutions. The first is run by the Canadian Mathematical Society. The second is run by a group called the Art of Problem Solving. All of the IMO problems from the beginning of the IMO in 1959 can be found in one or other of these sites but only some of the solutions are there. I think that this is a pity. Nevertheless you should find the sites useful. Happy problem solving and good luck to you for the future both within mathematics and without it.

Exercises

74. IMO 2008 Question 1. An acute-angled triangle ABC has orthocentre H. The circle passing through H with centre the midpoint of BC intersects the

line BC at A_1 and A_2. Similarly, the circle passing through H with centre the midpoint of CA intersects the line CA at B_1 and B_2, and the circle passing through H with centre the midpoint of AB intersects the line AB at C_1 and C_2. Show that A_1, A_2, B_1, B_2, C_1, C_2 lie on a circle.

(Proposed by Andrey Gavrilyuk, Russia)

75. IMO 2008 Question 2. (a) Prove that

$$\frac{x^2}{(x-1)^2} + \frac{y^2}{(y-1)^2} + \frac{z^2}{(z-1)^2} \geq 1$$

for all real numbers x, y, z, each different from 1, and satisfying $xyz = 1$.
(b) Prove that equality holds above for infinitely many triples of rational numbers x, y, z, each different from 1, and satisfying $xyz = 1$.

(Proposed by Walther Janous, Austria)

76. IMO 2008 Question 3. Prove that there exist infinitely many positive integers n such that $n^2 + 1$ has a prime divisor which is greater than $2n + \sqrt{2n}$.

(Proposed by Kęstutis Česnavičius, Lithuania)

77. IMO 2008 Question 4. Find all functions $f: (0, \infty) \to (0, \infty)$ (so, f is a function from the positive real numbers to the positive real numbers) such that

$$\frac{(f(w))^2 + (f(x))^2}{f(y^2) + f(z^2)} = \frac{w^2 + x^2}{y^2 + z^2}$$

for all positive real numbers w, x, y, z, satisfying $wx = yz$.

(Proposed by Hojoo Lee, South Korea)

78. IMO 2008 Question 5. Let n and k be positive integers with $k \geq n$ and $k - n$ an even number. Let $2n$ lamps labelled $1, 2, \ldots, 2n$ be given, each of which can be either *on* or *off*. Initially all the lamps are off. We consider sequences of *steps*: at each step one of the lamps is switched (from on to off or from off to on).

Let N be the number of such sequences consisting of k steps and resulting in the state where lamps 1 through n are all on, and lamps $n + 1$ through $2n$ are all off.

Let M be the number of such sequences consisting of k steps, resulting in the state where lamps 1 through n are all on, and lamps $n + 1$ through $2n$ are all off, but where none of the lamps $n + 1$ through $2n$ is ever switched on.

Determine the ratio N/M.

(Proposed by Bruno Le Floch and Ilia Smilga, France)

79. IMO 2008 Question 6. Let $ABCD$ be a convex quadrilateral with $|BA| \neq |BC|$. Denote the incircles of triangles ABC and ADC by ω_1 and ω_2 respectively. Suppose that there exists a circle ω tangent to the ray BA beyond A and to the ray BC beyond C, which is also tangent to the lines AD and CD. Prove that the common external tangents of ω_1 and ω_2 intersect on ω.

    (Proposed by Vladimir Shmarov, Russia)

80. IMO 2010 Question 4. Let P be a point interior to triangle ABC (with $CA \neq CB$). The lines AP, BP and CP meet its circumcircle Γ again at K, L, and M, respectively. The tangent line at C to Γ meets the line AB at S. Show that if $MK = ML$, then $SC = SP$.

    (Proposed by Poland)

8.14. Solutions

1. (i) From equation (1) we know that
    $$58 = k(2r + k - 1).$$
    What are the divisors of 58? Surely they are 1, 2, 29, 58.

    If $k = 1$, $2r + k - 1 = 58$. Hence $r = 29$. This solution is no good to us because it has $k = 1$.

    If $k = 2$, $2r + k - 1 = 29$. Hence $r = 14$. This gives us $14 + 15 = 29$. We've found one solution.

    If $k = 29$, $2r + k - 1 = 2$. Hence $r = -13$. Hang on, we said "sum of *positive* integers". No solutions here.

    If $k = 58$, $2r + k - 1 = 1$. In this case $r = -28$. Again no solutions.

    Why should we have known that $k = 29, 58$ could have been ignored?

    (ii) From equation (1) we have
    $$48 = k(2r + k - 1).$$
    The factors of 48 are 1, 2, 3, 4, 6, 8, 12, 16, 24, 48. These are all candidates for k. Or are they? Well, we can forget about $k = 1$. Why can we forget about $k = 8, 12, 16, 24, 48$? (Look at the factors of the right-hand side of equation (1).)

    So we have four possibilities.

    If $k = 2$, $2r + k - 1 = 24$. Hence $r = 11\,1/2$ (!!).

    If $k = 3$, $2r + k - 1 = 16$. Hence $r = 7$. So $7 + 8 + 9 = 24$ as required.

    If $k = 4$, $2r + k - 1 = 12$. Again r is not an integer.

    If $k = 6$, $2r + k - 1 = 8$. And again r is a fraction.

    So $k = 3$. There is only one way of writing 24 as the sum of consecutive positive integers.

(iii) We have to solve
$$32 = k(2r + k - 1).$$
No matter which factors of 32 we choose for k (other than 1) we get no solution. Why?

(iv) The factors of 144 are 1, 2, 3, 4, 6, 8, 9, 12, 16, 18, 24, 36, 48, 72, 144. We don't have to worry about $k = 1, 16, 18, 24, 36, 48, 72, 144$. This is because $k = 1$ is always out and the other k are **bigger** than $2r + k - 1$, the other factor.

If $k = 2$, $2r + k - 1 = 72$ and r is not an integer.

If $k = 3$, $2r + k - 1 = 48$ and $r = 23$.

If $k = 4$ then r is not an integer.

If $k = 6$ or 8 or 12 the same is true.

If $k = 9$, $2r + k - 1 = 16$ and $r = 4$.

2. (i) 1; (ii) 1; (iii) 0;

  (iv) 2 (here $72 = 23 + 24 + 25 = 4 + 5 + 6 + 7 + 8 + 9 + 10 + 11 + 12$).

3. We have seen that 16 behaves in this way. What's so special about 16?

  Let's have a look at the right hand side of equation (1). The two factors are k and $2r + k - 1$. Clearly if k is odd then $2r + k - 1$ is even. On the other hand if k is even, $2r + k - 1$ is odd.

  But you should also have realised by now that $2r + k - 1 > k$. So k is the smaller factor of k and $2r + k - 1$.

  So what numbers $2n$ have **only** the solution $k = 1$ to equation (1)? They have to be numbers with only one odd factor, otherwise there would be another solution for k. So $2n$ has to be a power of 2. Hence $n = 2^a$ for some $a \geq 0$.

  (Incidentally, if $n = 2^a$, then equation (1) has only one solution for k. Hence n cannot be written as the sum of (more than one) consecutive positive integer if and only if $n = 2^a$.)

4. In Exercise 2 we saw two numbers that can be written in only one way. One was 29, the other was 24. Does this give us any clues? Now 29 is a prime. But of course 24 isn't.

  Suppose n can only be written in one way.

  Case 1. n odd. Let $n = ab$, where $1 < a < b$ (and a, b are odd). By equation (1)
$$2ab = k(2r + k - 1).$$
  If $k = 2$, then $2r + k - 1 = ab$. Hence $2r = ab - 1$. This clearly has a solution.

  If $k = a$, then $2r + k - 1 = 2b$. Hence $2r = 2b - a + 1$. But $a < b$ so $2r > 0$. Further $2b - a + 1$ is even. We thus have a second solution.

But n can only be written as the sum in **one** way. Hence n is not the product of two odd numbers larger than 1. Hence n is prime.

Case 2. n even. Let $n = 2^c ab$ where $c \geq 1$ and $1 < a < b$, where a, b are odd.

By equation (1)

$$2^{c+1} ab = k(2r + k - 1).$$

If $k = a$, then $2r + k - 1 = 2^{c+1}b$. Hence $2r = 2^{c+1}b - a + 1$. This has a solution.

If $2^{c+1} < ab$, then $k = 2^{c+1}$ and $2r + k - 1 = ab$ has a solution. On the other hand if $2^{c+1} > bc$, then $k = bc$ and $2r + k - 1 = 2^{c+1}$ has a solution.

But n can only be written as a sum in **one** way. So $n = 2^c p$ where $c \geq 1$ and p is an odd prime.

You should now be able to prove the following theorem.

Theorem. *The number n can be written as the sum of consecutive positive integers in precisely one way if and only if $n = 2^m p$, where $m \geq 0$ and p is an odd prime.*

5. You should be thinking of a conjecture along the lines of "blah if and only if $n = 2^m pq$ where $m \geq 0$ and p, q are odd primes". Can $p = q$? Can we have $p \neq q$?!

6. How does 39 look?

The following values of k give a solution: $3, 3^2, 3^3, 3^4, 5, 3.5, 3^2.5, 3^3.5,$ $3^4.5, 5^2, 3.5^2, 5^3, 3^2.5^2, 3^3.5^2, 3^4.5^2, 5^4, 3.5^4, 3^2.5^4, 3^3.5^4, 3^4.5^4, 5^5$ as well as 2 and seventeen other even numbers. Can you find a formula for the number of sums if $n = 3^a 5^b$?

Remember, we're looking for the smaller of the factors cd in each case, where $cd = 2.3^4.5^7$. (Don't forget to discard $c = 1$.)

7. (a) (i) 6;   (ii) 12;   (iii) $54 = 9 \times 6$.
   (b) (i) 3;   (ii) 6;   (iii) 27.

8. (i) 2;   (ii) 5;   (iii) 26.
   Remember to discard the factor 1.

9. First of all recall that we need to factorise $2n$ in equation (1). There are $2(\alpha+1)(\beta+1)(\gamma+1)$ factors of $2n$. This is because we can choose to use 2^0 or 2; $p^0, p^1, p^2, \ldots, p^{\alpha-1}$ or p^α; $q^0, q^1, \ldots, q^{\beta-1}$ or q^β; $r^0, r^1, \ldots, r^{\gamma-1}$ or r^γ. We have 2 choices for the power of 2, $\alpha+1$ choices for the power of p, $\beta+1$ choices for the power of q and $\gamma+1$ choices for the power of r.

But factors come in pairs, a smaller together with a larger multiply together to give $2n$. Hence there are $(\alpha+1)(\beta+1)(\gamma+1)$ smaller factors.

All these smaller factors except the factor 1, are candidates for k. Hence the number of possible solutions is $(\alpha + 1)(\beta + 1)(\gamma + 1) - 1$.

10. How many legitimate factors for k are there. Remember we're looking for factors s, t such that $2n = st$, s is less than t and one of s, t is odd. (We then, of course, have to subtract the $k = 1$ possibility.)

    So we're looking at $2^{\alpha+1}q^\beta r^\gamma = st$. Since only one of s, t can be even, the $2^{\alpha+1}$ must be all in s or all in t. The number of factors of $2^{\alpha+1}q^\beta r^\gamma$ with $2^{\alpha+1}$ in one factor or the other is the same as the factors in $2q^\beta r^\gamma$. This is $2(\beta + 1)(\gamma + 1)$. Altogether, then, there are $(\beta + 1)(\gamma + 1)$ "smaller factors", one of which is one. So the number of consecutive sums is $(\beta + 1)(\gamma + 1) - 1$.

11. The answer is $(\alpha_1 + 1)(\alpha_2 + 1) \cdots (\alpha_r + 1) - 1$ if p_1 is odd and $(\alpha_2 + 1)$ $(\alpha_3 + 1) \cdots (\alpha_r + 1) - 1$ if $p_1 = 2$.

12. We know from Problem 1 that 9 will do. But what is the general situation? By the last exercise we're looking for $n = p_1^2$ or $2^{\alpha_1}p_2^2$.

    Incidentally, the only two consecutive sums are

    (i) for $n = p_1^2$, $\frac{p_1^2-1}{2} + \frac{p_1^2+1}{2}$ and $\frac{p_1+1}{2} + \frac{p_1+3}{2} + \cdots + \frac{3p_1-1}{2}$; and

    (ii) for $n = 2^{\alpha_1}p_2^2$, $\frac{2p_2^2-1}{2} + \frac{2p_2^2+1}{2}$ and $\frac{3p_1+1}{2} + \frac{3p_1+3}{2} + \cdots + \frac{5p_1-1}{2}$.

13. Suppose n is odd. Then $(\alpha_1 + 1)(\alpha_2 + 1) \cdots (\alpha_r + 1) - 1 = 4$. This forces $\alpha_1 = 4$, so $n = p_1^4$.

    Suppose n is even. Then $(\alpha_2 + 1) \cdots (\alpha_r + 1) - 1 = 4$, so $\alpha_2 = 4$ and $n = 2^{\alpha_1}p_2^4$.

14. Whether n is even or odd, equation (1) gives $2n = 1990(2n + 1989)$. So $n = 995(2n + 1989)$. This forces n to be odd. Since $995 = 5 \times 199$, then this means that $n = 5^{\alpha_1}199^{\alpha_2}p_3^{\alpha_3} \ldots p_r^{\alpha_r}$.

    But $(\alpha_1 + 1)(\alpha_2 + 1) \cdots (\alpha_3 + 1) = 1991 = 11 \times 181$. So $\alpha_1 = 10$, $\alpha_2 = 180$ and $\alpha_i = 0$ for $i \geq 2$ or $\alpha_1 = 180$, $\alpha_2 = 10$ and $\alpha_i = 0$ for $i \geq 2$. Hence $n = 5^{10}199^{180}$ or $5^{180}199^{10}$.

15. What did you manage to find?

16. This recurrence relation is awfully close to the Fibonacci relation. It looks as if it might be worth trying $b_n = na_n$. Then we get $b_0 = 0 \cdot a_1 = 0$ and $b_1 = 1 \cdot a_1 = 1$. Further $b_{n+1} = b_n + b_{n-1}$.

    The quick method of solution here is via $x^2 = x + 1$. So $x = \frac{1 \pm \sqrt{5}}{2}$.

    Let $\alpha = \frac{1+\sqrt{5}}{2}$ and $\alpha = \frac{1-\sqrt{5}}{2}$. Then $b_n = K_1\alpha^n + K_2\beta^n$. To find K_1 and K_2 we note that

    $$b_0 = 0 = K_1 + K_2$$

    and

    $$b_1 = 1 = K_1\alpha + K_2\beta.$$

So

$$1 = K_1(\alpha - \beta) = K_1\sqrt{5}.$$

Hence

$$b_n = \frac{1}{2^n\sqrt{5}}\left[(1+\sqrt{5})^n - (1-\sqrt{5})^n\right].$$

Finally

$$a_n = \frac{1}{n2^n\sqrt{5}}\left[(1+\sqrt{5})^n - (1-\sqrt{5})^n\right].$$

17. Let $b_n = 2^n a_n$. Then $b_0 = 1$, $b_1 = 2$ and $b_{n+1} = b_n + b_{n-1}$. Hence $b_n = K_1\alpha^n + K_2\beta^n$, where α and β are as in the last exercise.
    So

$$1 = K_1 + K_2,$$

and

$$2 = K_1\alpha + K_2\beta.$$

Hence

$$K_1 = \frac{2-\beta}{\alpha-\beta} = \frac{3+\sqrt{5}}{2\sqrt{5}} \quad \text{and} \quad K_1 = \frac{\sqrt{5}-3}{2\sqrt{5}}.$$

So

$$a_n = \frac{1}{2^{n+1}\sqrt{5}}\left\{(3+\sqrt{5})\left(\frac{1+\sqrt{5}}{2}\right)^n + (\sqrt{5}-3)\left(\frac{1-\sqrt{5}}{2}\right)^n\right\}.$$

18. You should be getting the idea now. Here try $g(n) = 2^n f(n)$, so $g(0) = 1$, $g(1) = 2$ and $g(n+1) = g(n) - g(n-1)$. The appropriate equation here is $x^2 = x-1$. This has solution $\alpha = \frac{1+i\sqrt{3}}{2}$ and $\beta = \frac{1-i\sqrt{3}}{2}$. Now that's alright if you know about $i = \sqrt{-1}$. But if that scares you, is there something else that you can do? Try finding $g(0), g(1), g(2), \ldots, g(16)$. Can you see any pattern?
    Why is $g(6n+2) = g(6n) = -g(6n+3) = -g(6n+5) = 1$ and $g(6n+1) = 2 = -g(6n+4)$? So now you can find $f(n)$?

19. $f(n+2) = 2^{2n+4}f(n+1) - 2^{4n+4}f(n) + n2^{n^2}$. I guess those last three Exercises weren't all that much help after all. It's not as easy as they were.

20. If you take the hint, you'll get

$$2^{h(n+2)}g(n+2) = 2^{2(n+2)}2^{h(n+1)}g(n+1) - 2^{4(n+1)}2^{h(n)}g(n).$$

Now what can we guess for $h(n)$ that will make life easier? Can we get the multiples of $g(n+2)$, $g(n+1)$ and $g(n)$ down to constants?

After some experimentation, you might try $h(n) = n^2$. Then

$$2^{n^2+4n+4}g(n+2) = 2^{2n+4}2^{n^2+2n+1}g(n+1) - 2^{4n+4}2^{n^2}g(n).$$

This reduces to $g(n+2) = 2g(n+1) - g(n)$, which is a little recurrence relation that is eminently solvable.

21. So what happens to equation (2) with $f(n) = 2^{n^2}g(n)$? Surely it becomes

$$g(n+2) = 2g(n+1) - g(n) + n2^{n^2}2^{-n^2-4n-4}$$
$$= 2g(n+1) - g(n) + n16^{-n-1}.$$

But what on earth are we going to do with the $n16^{-n-1}$?

22. It's perhaps still a bit mysterious like that, although it can be done. Rewrite the equation like this, though.

$$g(n+2) - 2g(n+1) + g(n) = 0,$$
$$g(n+1) - 2g(n) + g(n-1) = 0,$$
$$g(n) - 2g(n-1) + g(n-2) = 0,$$
$$\cdots$$
$$g(4) - 2g(3) + g(2) = 0,$$
$$g(3) - 2g(2) + g(1) = 0,$$
$$g(2) - 2g(1) + g(0) = 0.$$

Now you will note that all the $g(2)$'s cancel, as do all the $g(3)$'s, $g(4)$'s, $\ldots, g(n-2)$'s, $g(n-1)$'s and $g(n)$'s. All we've got left is

$$g(n+2) - g(n+1) - g(1) + g(0) = 0.$$

But

$$g(1) = 0 = g(0).$$

So

$$g(n+2) = g(n+1).$$

What this means, of course, is that $g(n) = 0$.

23. Repeating the method of the last exercise we get

$$g(n+2) = g(n+1) + [1 + 1 + \cdots + 1] = g(n+1) + n + 1.$$

To get on top of this, repeat the trick of the last exercise.

So

$$g(n+2) - g(n+1) = n+1,$$

$$g(n+1) - g(n) = n,$$

$$\cdots$$

$$g(3) - g(2) = 2,$$

$$g(2) - g(1) = 1.$$

Hence $g(n+2) - g(1) = (n+1) + n + \cdots + 3 + 2 + 1$.

So $g(n+2) = 1/2(n+1)(n+2)$. If you like, then $g(n) = 1/2n(n-1)$.

24. Here $g(n+2) = g(n+1) + [(n+2) + (n+1) + \cdots + 3 + 2]$. So

$$g(n+2) - g(n+1) = 1/2(n+1)(n+4),$$

$$g(n+1) - g(n) = 1/2n(n+3),$$

$$\cdots$$

$$g(3) - g(2) = 1/2 \cdot 2 \cdot 5,$$

$$g(2) - g(1) = 1/2 \cdot 1 \cdot 4.$$

So $g(n+2) = 1/2[(n+1)(n+4) + n(n+3) + \cdots + 2 \cdot 5 + 1 \cdot 4]$.
Actually, we've still got a small problem on the right-hand side of the equation. But it's only a small problem. A closer look will give you

$$g(n+2) = \frac{1}{2} \sum_{r=1}^{n+1} r(r+3) = \frac{1}{2} \left[\sum_{r=1}^{n+1} r^2 + 3 \sum_{r=1}^{n+1} r \right]$$

$$= \frac{1}{2} \left[\frac{1}{6}(n+1)(n+2)(2n+3) + \frac{3}{2}(n+1)(n+2) \right]$$

$$= \frac{1}{6}(n+1)(n+2n+6),$$

where I've skipped a little algebra. So actually,

$$g(n) = 1/6(n-1)n(n+4).$$

25. The method is clear. The algebra is a pain but maybe you have CAS handy.

26. $g(n+2) - 2g(n+1) + g(n) = n16^{-n-1}$ reduces to

$$g(n+2) - g(n+1) = n16^{-n-1} + (n-2)16^{-n+1} + \cdots + 0.$$

The problem now seems to be, how on earth can we sum

$$n16^{-n-1} + (n-1)16^{-n} + (n-2)16^{-n+1} + \cdots + 0?$$

27. Here $r = 16^{-1}$, so $\sum_{t=0}^{n} 16^{-t} = \frac{16^{-n}-1}{16^{-1}-1} = \frac{16^{n}-1}{15 \cdot 16^{n-1}}$.

28. Those of you who have not yet met calculus should ask your teacher for a book to work on. For the rest, let $y = \sum_{t=0}^{n} r^t$. Now $\frac{dy}{dr} = \sum_{t=0}^{n} t r^{t-1}$. But we also know that $y = \frac{r^{n+1}-1}{r-1}$, so, by the quotient rule $\frac{dy}{dx} = \frac{nr^{n+1}-(n+1)r^n+1}{(r-1)^2}$.

Hence $\sum_{t=0}^{n} t r^{t-1} = \frac{nr^{n+1}-(n+1)r^n+1}{(r-1)^2}$.

29. From the last exercise with $r = 16$ we get
$$\sum_{t=0}^{n} t 16^{t-1} = \frac{n16^{n+1} - (n+1)16^n + 1}{15^2}.$$

30. We need to start from $\sum_{t=0}^{n} r^{-t}$. Let $y = \sum_{t=0}^{n} r^{-t}$. Then
$$\frac{dy}{dt} = \sum_{t=0}^{n} -t r^{-t-1}.$$

On the other hand $y = \frac{r^{-n-1}-1}{r^{-1}-1}$, so using the quotient rule we get $\frac{dy}{dt} = \frac{-nr^{-n-1}+(n+1)r^{-n}-1}{(1-r)^2}$. Hence
$$\sum_{t=0}^{n} -t r^{-t-1} = \frac{-nr^{-n-1} + (n+1)r^{-n} - 1}{(1-r)^2}.$$

31. From the last exercise $\sum_{t=0}^{n} t 16^{-t-1} = \frac{n16^{-n-1}-(n+1)16^{-n}+1}{15^2}$.

32. Differentiate $\sum_{t=0}^{n} r^{-t}$ twice.

33. The last time we saw Problem 2 was in Exercise 26. There we had
$$g(n+2) - g(n+1) = \sum_{t=0}^{n} t 16^{-t-1}.$$

From Exercise 31 we now know that
$$g(n+2) - g(n+1) = \frac{n16^{-n-1} - (n+1)16^{-n} + 1}{15^2}.$$

It might help if we change this to
$$g(n+1) - g(n) = \frac{(n-1)16^{-n} - n16^{-n-1} + 1}{15^2}$$

If we now add $g(n+1) - g(n)$ to $g(n) - g(n-1)$ to $g(n-1) - g(n-2)$, etc., in the usual way, we get
$$g(n) = \frac{1}{15^3}[15n - 32 + (15n+2)16^{-n+1}].$$

So
$$f(n) = \frac{2n^2}{15^3}[15n - 32 + (15n+2)16^{-n+1}]$$
$$= \frac{2(n-2)^2}{15^3}[(15n-32)16^{n-1} + 15n + 2].$$

Now using the fact that $1990 \equiv 1 \pmod{13}$, $3^3 \equiv 1 \pmod{13}$, and $1990 \equiv 1 \pmod{13}$, it can be shown that $f(1989) \equiv 0 \pmod{13}$, $f(1990) \equiv 0 \pmod{13}$ and $f(1991) \equiv 0 \pmod{13}$.

34. (a) To find $f(1)$ should we put $x = 1$ or $y = 1$?

Try $x = 1$. Then $f(f(y)) = \frac{f(1)}{y}$. Gulp!

Try $y = 1$. Then $f(xf(1)) = f(x)$. Well $f(1) = 1$ would be consistent but $f(xf(1)) = f(x)$ doesn't imply $f(1) = 1$ unless f is a 1:1 function (sometimes called an ***injection***). In other words if $f(x_1) = f(x_2)$ then $x_1 = x_2$. So can we show that f is an injection? What about $f(2)$?

Try $x = 2$. Then $f(2f(y)) = \frac{f(2)}{y}$. Gulp (again)!

Try $y = 2$. Then $f(xf(2)) = \frac{f(x)}{2}$. That doesn't even give us room for a decent guess. All we've discovered is that the image of some multiple of x is a half of the image of x.

What about $f(3)$? A bit of experimenting doesn't get us any further than $f(2)$.

(b) A lot of these problems end up $f(x) = x$ or something relatively simple. Is that the case here?

If $f(x) = x$, then $f(xf(y)) = xf(y) = xy$. Similarly $\frac{f(x)}{y} = \frac{x}{y}$. Going back to the functional equation we get $xy = \frac{x}{y}$, so $y = \frac{1}{y}$. Hmm. That's not going to work for many values of y! So $f(x) \neq x$ in general.

How about $f(x) = \frac{1}{y}$? Then $f(xf(y)) = \frac{1}{xf(y)} = \frac{y}{x}$, while $\frac{f(x)}{y} = \frac{1}{xy}$. Blast! Again that only works if $y = \frac{1}{y}$. What other reasonable relations are there?

35. Remember, if $f(x_1) = f(x_2)$ implies $x_1 = x_2$, then f is injective.

(i) Suppose $f(x_1) = f(x_2)$. Then $x_1 = x_2$, since $f(x_1) = x_1$ and $f(x_2) = x_2$. So $f(x) = x$ is an injection.

(ii) Suppose $f(x_1) = f(x_2)$. Then $x_1^2 = x_2^2$. But this gives $x_1 = \pm x_2$. So $x_1 = x_2$ is not forced. Hence $f(x) = x^2$ is not an injective function.

(iii) Suppose $f(x_1) = f(x_2)$. Then $x_1^3 = x_2^3$.

Hence $x_1 = x_2$ and $f(x) = x^3$ is an injection.

36. Actually $f: R^+ \to R^+$, where $f(x) = x^2$ ***is*** 1:1. This is because if $f(x_1) = f(x_2)$, then $x_1^2 = x_2^2$. But in R^+ there is only one solution to this. That's $x_1 = x_2$.

Another 1:1 function is $f: R^+ \to R^+$, $f(x) = e^x$. Try the test.

If $f(x_1) = f(x_2)$, then $e^{x_1} = e^{x_2}$.

Hence $e^{x_1 - x_2} = 1$. So $x_1 - x_2 = 0$. This gives $x_1 = x_2$.

37. Suppose $f(x_1) = f(x_2)$. Then $f(x_1 f(y)) = \frac{f(x_1)}{y} = \frac{f(x_2)}{y} = f(x_2 f(y))$. But how does $f(x_1 f(y)) = f(x_2 f(y))$ help?

Suppose $f(y_1) = f(y_2)$. Then $f(x f(y_1)) = \frac{f(x)}{y_1}$ and $f(x f(y_2)) = \frac{f(x)}{y_2}$. But $f(y_1) = f(y_2)$, so $f(x f(y_1)) = f(x f(y_2))$. That means that $\frac{f(x)}{y_1} = \frac{f(x)}{y_1}$. Hence $\frac{1}{y_1} = \frac{1}{y_1}$. So $y_1 = y_2$. Done it!

38. $f(1 f(2)) = \frac{f(1)}{2}$. Since $f(1) = 1$, $f(f(2)) = \frac{1}{2}$. Now let $x = 1$, $y = 1/2$, then $f(f(1/2) = 2$. I suppose we're now trying to see if $f(f(y)) = 1/y$ and $f(f(1/y)) = y$. But these are both obviously true. Just put $x = 1$, $y =$ "y" or $x = 1$, $y =$ "$1/y$". So have we got the beginning of a cycle?

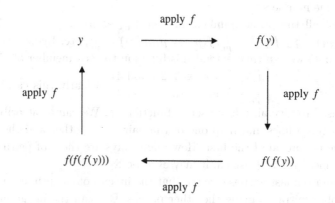

39. Guess. Suppose $f(2) = 5$. Then what is $f(5)$? Clearly $f(5) = f(f(2)) = 1/2$. So what is $f(1/2)$? $f(1/2) = f(f(5))$. But we know about $f(f(y))$. It's $1/y$. So $f(f(5)) = 1/5$. So if $f(2) = 5$,

40.

$$2 \xrightarrow{f} 5 \xrightarrow{f} \tfrac{1}{2} \xrightarrow{f} 1/5 \xrightarrow{f} 2$$

$$2 \xrightarrow{f} a \xrightarrow{f} \tfrac{1}{2} \xrightarrow{f} 1/a \xrightarrow{f} 2$$

41.

$$b \xrightarrow{f} c \xrightarrow{f} 1/b \xrightarrow{f} 1/c \xrightarrow{f} b$$

42. $2 \to 5 \to 1/2 \to 1/5$. Then suppose $10 \to 6 \to 1/10 \to 1/6$. Are there any problems with this? Now $f(10) = f(2 \times 5) = f(2f(2)) = f(2)/2 = 5/2$. So given $f(2) = 5$, we are restricted in some way as to the choice of $f(10)$. What is the restriction?

43. $f(2 \times 5) = 5/2$ as we have seen. $f(2)f(5) = 5 \times 1/2 = 5/2$. Is f a multiplicative function? Consider $f(ab)$. Suppose c is such that $f(c) = b$.

Then $f(ab) = f(af(c)) = f(a)/c$. But we know that we have the cycle $c \to b \to 1/c \to 1/b$. So $f(b) = 1/c$. Hence $f(ab) = f(a)/c = f(a)f(b)$. So f is a multiplicative function.

44. $f(n) = f(p_1^{\alpha_1} p_2^{\alpha_2} \dots p_r^{\alpha_r}) = f(p_1^{\alpha_1}) f(p_2^{\alpha_2}) \dots f(p_r^{\alpha_r})$
    $= f(p_1)^{\alpha_1} f(p_2)^{\alpha_2} \dots f(p_r)^{\alpha_r}$.

Hence once the image of all the prime numbers is defined, so is the image of **every** positive integer.

45. $f(\frac{m}{n}) = f(m)f(\frac{1}{n}) = \frac{f(m)}{f(n)}$. So again, once the image of the positive integers is defined then so are all the rationals in Q^+. Consequently the image of all the members of Q^+ depends solely on the image of the prime numbers.

46. List all the prime numbers in order, $p_1 < p_2 < p_3 < \cdots$. Then let $f(p_1) = p_2, f(p_2) = \frac{1}{p_1}, f(p_3) = p_4, f(p_4) = \frac{1}{p_3}$, etc. Using Exercises 44 and 45 we can then extend this function to each member of Q^+.

47. Let $f(p_i) = \begin{cases} p_{i+2}, & \text{for } i \equiv 1, 2 \pmod 4, \\ \frac{1}{p_{i-2}}, & \text{for } i \equiv 3, 0 \pmod 4. \end{cases}$ Clearly this can be generalised to give an infinite set of functions f. We can arbitrarily pair up all the primes and map one of the pair to the other and the other to the reciprocal of the first. How many ways are there of pairing up the primes? It's at least $\aleph_0$, it may even be $\aleph_1$.

48. From the last exercise we see that the images of the primes are reasonably arbitrary among the other primes. But can the image of a prime number be composite or rational (not equal to a prime or the reciprocal of a prime)? Do we get into any trouble with the following functions?

$$p_1 \to p_2 p_3 \to 1/p_1 \to 1/p_2 p_3 \to p_1$$
$$p_2 \to p_4/p_1 \to 1/p_2 \to p_1/p_4 \to p_2$$
$$p_3 \to 1/p_4 \to 1/p_3 \to p_4 \to p_3$$
$$p_5 \to p_6 \to 1/p_5 \to 1/p_6 \to p_5$$
$$p_7 \to p_8 \to 1/p_7 \to 1/p_8 \to p_7$$
$$\cdots$$
$$p_i \to p_{i+1} \to 1/p_i \to 1/p_{i+1} \to p_i \quad \text{etc.}$$

It looks as if there are an incredible number of solutions to this functional equation.

50. Using the techniques above it is easy to show that f is a 1:1 function and that $f(1) = 1$.

Let $x = 1$. Then $f(f(y)) = y$. This is clearly satisfied by $f(t) = t$ for all $t \in Q^+$.

Ah, but there is at least one other very simple function with the property that $f(f(y)) = y$. Are there more than one? You mean to say there are an infinite number of these too?

51. From Chapters 2 and 3 of *First Step*, you should know that if the edges of K_6 are arbitrarily coloured in one of two colours then there is a monochromatic triangle. Without loss of generality suppose this triangle is coloured red. Surely there's a K_6 inside any K_{10}. Hence our K_{10} contains a red triangle.

52. Remove the red triangle from K_{10} and you've at least got a K_7 left. This has to have a monochromatic triangle. If the monochrome is red we're finished. Unfortunately it might be blue.

53. No. We have at least a K_4 left and it is possible to colour its edges in two colours so that there are no monochromatic triangle.

    Does K_4 have a monochromatic odd cycle? Not necessarily.

54. From Exercise 12 in Chapter 2 of *First Step*, we know that we cannot guarantee a monochromatic triangle. Can we guarantee a monochromatic 5-cycle? Check out the proof below.

    Lemma. *Arbitrarily colour the edges of K_5 either red or blue. Then K_5 either contains a monochromatic triangle or a monochromatic 5-cycle.*
    Proof. (Try to prove this for yourself before you cheat and see what I did.) Consider the colours at the vertex a. Without loss of generality we may assume that there are two, or at least three, edges coloured red.
    Case 1. Suppose at least three edges are coloured red. Then the edges ab, ac, ad, say, are red. If there is a red edge joining any of b, c, d, we get a red triangle. If all of these edges are blue we get a blue triangle. This gives us our monochromatic triangle.
    Case 2. So we may suppose that at every vertex precisely two edges are red (and precisely two are blue). Otherwise we can use the argument of the last paragraph to force a monochromatic triangle. Hence ab, ac are red (without loss of generality.) There must be another red edge at b by the assumption of this case. If it is bc we have a red triangle. If not, suppose bd is red.

    Now cd is not red because then the two red edges at e have nowhere to go. So ce is red. Hence de is red and we have a red 5-cycle. (Incidentally, we simultaneously have a blue 5-cycle as well.)   □

55. So far we have two disjoint triangles. One red, one blue. Suppose the red one is on vertices u, v, w and the blue one on x, y, z.

    Suppose edges ux, uy are blue. Then u, x, y form a blue triangle. This blue triangle or the red one on u, v, w, along with the monochromatic triangle or 5-cycle of the lemma of the last exercise give the result we want. (This lemma can be applied to the four unlabelled vertices plus z.)

So at most one blue edge goes from u to x, y, z. Suppose ux, uy are red. If vx is red, then u, v, x is a red triangle and we finish the problem by the argument of the last paragraph. If vx is blue, then look at vy. If it's red, then apply the lemma to w and the four other vertices (since u, v, y is now red). If it's blue, then v, x, y is blue and we apply the lemma to z and the four other vertices.

However you colour it you get two disjoint odd cycles in the same colour. End of Problem 4. (Actually there is a neater argument than this. Find it.)

56. The problem is that Exercise 13 doesn't guarantee us *disjoint* triangles which are both red, both blue or one of each.

    Colour K_5 all red. Add a sixth vertex. Join the new vertex to K_5 by blue edges. Thus K_6 has no pair of disjoint red triangles or blue triangles or one red and one blue triangle.

57. According to JW Moon, Disjoint triangles in chromatic graphs, Mathematics Magazine, Vol. 39, No. 5 (Nov. 1966), pp. 259–261, the maximum number of disjoint monochromatic triangles in K_n is $\lfloor \frac{n}{3} \rfloor - 1$ or $\lfloor \frac{n}{3} \rfloor$. Further, if $n \equiv 2 \pmod 3$ and $n \geq 8$, then the number is $\lfloor \frac{n}{3} \rfloor$. This tells us that K_8 has two disjoint monochromatic triangles and K_7 has 1 or 2. So is the best possible number here 7 or 8? It shouldn't take long to see if K_7 can have two disjoint monochromatic triangles.

58. Of course $n \geq 8$. But it's hardly likely to be that small.

59. This looks like a nice problem for a wet weekend. (Or a computer!)

60. If $n = 1$, we get the number 7. This is prime so $n = 1$ satisfies Problem 5.

    If $n = 2$, we get 17 and 71. Both of these are prime. So $n = 2$ satisfies Problem 5. Doing fine.

    If $n = 3$, we get 117, 171, 711. Oh dear! These are all divisible by 3!

61. If $n = 4$, we get 1117, 1171, 1711, 7111. A bit of experimenting will show you that 1711 isn't prime. What about the others though?

    If $n = 5$, we get 11117, 11171, 11711, 17111, 71111. I think 11711 is divisible by 7. What about the others?

    If $n = 6$, aren't they all divisible by 3?

62. Well it looks as if when n is divisible by 3 then so are all the numbers proposed.

    Is it worth conjecturing at this stage that the only n for which all the proposed numbers are prime are $n = 1$ and $n = 2$? Perhaps its a little early for this guess.

63. Have a go at the $n = 3k$ conjecture.

Let N be any number consisting of $n-1$ ones and 1 seven. Let $n = 3k$. Then the sum of the digits of N is $3k - 1 + 7 = 3k + 6$. This sum is divisible by 3. Hence N is divisible by 3. (Have a look at Chapter 4 of *First Step* for a proof of this fact. Incidentally you should remember this little divisibility by 3 test.)

If we could prove the other conjecture, then we'd have solved Problem 5. Is it a bit early to worry about that yet?

64. The observation that needs to be made is that N is really of the form $111 \cdots 1$ (n ones) plus either 6 or 60 or 600 etc. So N can be written in the form $J_n + 6 \times 10^k$ where J_n is the n-digit number made up completely of ones and k is such that $0 \le k \le n - 1$.

65. That list looks larger than it needs to. Since the digits 1 and 7 are odd, no N is ever going to be divisible by 2, 4, 6, 8, 10 or any other even number for that matter.

We've already dealt with 3. Divisibility by 5 is never going to be possible since N is never going to end in 0 or 5. The only realistic possibilities in the list then, are 7 and 11. Let's try working on 7 because it is smaller than 11.

If $n = 4$, none of the N are divisible by 7. But at least one N is divisible by 7 for $n = 5$. Now try $n = 6$. Here
$$N = J_6 + 6 \times 10^k.$$
Ah! Now that's interesting J_6 is divisible by 7. Is there some $k, 0 \le k \le 5$ for which 6×10^k is divisible by 7.

$k = 0$: $6 \equiv 6 \pmod 7$, $\qquad k = 1$: $60 \equiv 4 \pmod 7$,

$k = 2$: $600 \equiv 5 \pmod 7$, $\qquad k = 3$: $6000 \equiv 1 \pmod 7$,

$k = 4$: $60000 \equiv 3 \pmod 7$, $\qquad k = 5$: $600000 \equiv 2 \pmod 7$.

Hum! That was a waste of time for two reasons. First 6×10^k is not divisible by 7 so none of the N are. Second, and much more important, $n = 6$ is divisible by 3. So I already knew I wouldn't find any primes there. It didn't matter too much then if N was divisible by 7 (as well as 3).

Let's try $n = 7$. Then
$$N = J_7 + 6 \times 10^k, \quad \text{for } 0 \le k \le 6.$$
rewriting we get
$$N = 10J_6 + 1 + 6 \times 10^k.$$
I took out the $10J_6$ since J_6 is divisible by 7.

Ah! Now we're in business. If $k = 0, 1 + 6$ is divisible by 7. Hence so is $N = 1111117$.

You should now be able to find an N which is divisible by 7 for $n = 8, 10, 11, 13$ and 14.

66. $n = 8$. $N = J_8 + 6 \times 10^k = 10^2 J_6 + J_2 + 6 \times 10^k$.

Now $J_6 \equiv 0 \pmod 7$ $J_2 = 11 \equiv 4 \pmod 7$. Can we find k such that $6 \times 10^k \equiv 3 \pmod 7$? Looking at the list in the last exercise we see that $k = 4$ will do. So $J_8 + 6 \times 10^4 = 11171111$ is divisible by 7.

$n = 10$. $N = J_{10} + 6 \times 10^k = 10^4 J_6 + J_4 + 6 \times 10^k$. Since $J_4 = 1111$ $\equiv 5 \pmod 7$, from the last Exercise we choose $k = 5$. Hence 1111711111 is divisible by 7.

$n = 11$. $N = J_{11} + 6 \times 10^k = 10^5 J_6 + J_5 + 6 \times 10^k$. Since $J_5 = 11111$ $\equiv 2 \pmod 7$, so we choose $k = 2$ to give 11111111711 is divisible by 7. (Actually we would have got this from Exercise 61. All we had to do was to remember that 11711 is divisible by 7. So $10^5 J_6 + 11711$ surely had to be too.)

$n = 13$. $N = J_{13} + 6 \times 10^k = 10^7 J_6 + 10 J_6 + J_1 + 6 \times 10^k$. Since $J_1 \equiv 1 \pmod 7$, so we choose $k = 0$ to give 1111111111117 is divisible by 7. (We could have got this from the $n = 7$ case done in the last exercise. The other thing to note here is that 1111111111117 is not the only number N for $n = 13$, which is divisible by 7.)

$n = 14$. $N = sJ_6 + J_2 + 6 \times 10^k$ for s some whole number. So take $k = 4$ to give 11111111117111. (Again (1) we could have got this immediately from $n = 8$ and (2) this is not the only candidate for divisibility by 7 honours.)

67. The m.o. (from the Latin *modus operandi* meaning method of operation) as they used to say in all good detective novels, is now clear. In general, $N = J_n + 6 \times 10^k$. If $n \equiv 0 \pmod 3$, then N is divisible by 3 and so is not prime. For $n \not\equiv \pmod 3$, then $n = 6m + r$, where $r = 1, 2, 4, 5$. Then
$$N = J_n + 6 \times 10^k = sJ_6 + J_r + 6 \times 10^k,$$
where s depends on m.

Suppose $m \geq 1$. If $r = 1$, choose $k = 0$. If $r = 2$, choose $k = 4$. If $r = 4$, choose $k = 5$. If $r = 5$, choose $k = 2$. Hence for $n = 6m + r$ with $m \geq 1$, N is divisible by 7.

We have already checked $n = 1, 2, 3, 4, 5, 6$ in Exercises 60 and 61. Hence we conclude that for a given n, N is always prime only if $n = 1, 2$.

68. A. What happens if 7 is changed to 3 in Problem 5?

B. What happens if 7 is changed to some other number?

C. What happens if 1 is changed to some other number?

D. What happens if 1 and 7 are changed to other numbers?

E. What happens if we use $n - 2$ ones and 2 sevens?

F. For what n in Problem 5 is there at least one N which is prime?

G. What happens if we combine all variations of D, E and F?

69. This must surely be solvable using the same approach we used to Problem 5. It might be slightly more interesting though because I think that 3; 13, 31; 113, 131, 311 are all prime. My guess is that there is at least one composite number for all $n \geq 4$. Can someone confirm that for me?

70. In some ways this is a far more interesting question than Problem 5. Of course that may be because at this stage I don't know the answer. Initially it does look difficult. But it may be that for some n sufficiently large, it is possible to show that all N are divisible by some integer t. Again, I'd be interested to hear what people find.

    Just to start you off, of course, there *is* at least one prime for $n = 1, 2$ and I'm sure you can work out the cases 4, 5, 7, 8, etc for yourselves.

    Naturally some numbers are not worth using. For instance, if we replace 7 in Problem 5 by 0, 5 or any even number we don't find too many n's with no prime N's. What about 9 though?

    If I were you I'd grab hold of a table of primes up to seven digits long. You'll be able to check conjectures a lot quicker that way.

71. How did you go on this one?

72. How did you go with this? If you found anything interesting let me know and we might be able to put it in the next edition of this book.

73. I hope that you found some interesting results here and noticed a few special cases that generalised.

74. Note here that there are at least three different solutions to this problem. I will only be considering one of these. So you may have a solution that doesn't look like the one that follows.

    Draw a diagram.

    Experiment with special triangles: right angled triangle; equilateral triangle; isosceles triangle. Do any of these special cases lead to any general ideas?

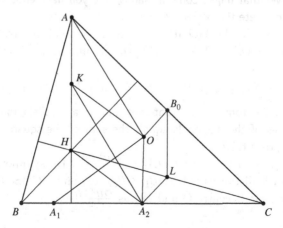

One approach here is to first try to identify the centre of the circle. If we can guess what that might be, we would then have to show that all of the points under consideration are an equal distance from that potential centre. Did your experimenting suggest a possible centre for the circle? What might be 'natural' centre points to try?

The perpendiculars of the segments A_1A_2, B_1B_2, C_1C_2 are also the perpendicular bisectors of BC, CA, AB. So they meet at O, the circumcentre of ABC. Thus O is the only point that can possibly be the centre of the desired circle.

Having identified the centre of the circle, all we have to do now is to show that all of the six given points are equidistant from that point. It's now a matter of chasing around the diagram and using symmetry to cut down the work needed. What lines are parallel? What line segments are equal in length? Where are there some useful triangles? Can you finish it off from here?

From the right angled triangle OA_2A_1, we get

$$OA_1^2 = OA_2^2 + A_2A_1^2 = OA_2^2 + A_2H^2. \tag{1}$$

Let K be the midpoint of AH and let L be the midpoint of CH. Since A_2 and B_0 are the midpoints of BC and CA, we see that $A_2L \parallel BH$ and $B_0L \parallel AH$. Thus the segments A_2L and B_0L are perpendicular to AC and BC, hence parallel to OB_0 and OA_2, repectively. Consequently OA_2LB_2 is a parallelogram, so that OA_2 and B_0L are equal and parallel. Also the midline B_0L of triangle AHC is equal and parallel to AK and KH.

What useful information does this give us? How does it get us to our goal of showing six distances are equal? What properties of the situation haven't we used yet that might come in handy? Do you have enough information now to complete the problem?

It follows that AKA_2O and HA_2OK are parallelograms. The first one gives $A_2K = OA = R$, where R is the circumradius of ABC. From the second one we obtain

$$2(OA_2^2 + A_2H^2) = OH^2 + A_2H^2 = OH^2 + R^2. \tag{2}$$

This comes from a result that says that, in a parallelogram, the sum of the squares of the diagonals equals the sum of the squares of the sides. Can you prove this?

From (1) and (2) we get $OA_1^2 = \frac{OH^2+R^2}{2}$. By symmetry, the same holds for the distances OA_2, OB_1, OB_2, OC_1 and OC_2. So the six points do lie on a circle centre O and radius $\frac{OH^2+R^2}{2}$.

75. It might be worth looking at the two variable case and taking z out of the equation. This might give you some ideas of how to proceed with the three variables. Does it look like an AM/GM problem? What other approaches might be taken?

    (a) Try starting with the substitution $\frac{x}{x-1} = a, \frac{y}{y-1} = b, \frac{z}{z-1} = c$.

    Play with that to see how far it will get you. Is it possible that this will lead to something being equal to a square? What might that something be if it shows what we are trying to show?

    So we now need to prove that $a^2+b^2+c^2 \geq 1$ subject to the constraints and $(a-1)(b-1)(c-1) = abc$.

    I'll leave you to do the algebra involved here. It should be straightforward. The graph of $r = \frac{s}{s-1}$ should convince you that $a \neq 1$ is the only restriction on a. I'll also leave you to do the work that will get you to the following equation. Again it shouldn't be too hard. Getting the idea to go in this direction is the hard part of (a).

    So
    $$a^2 + b^2 + c^2 - 1 = (a+b+c-1)^2$$
    and that gets us where we wanted to go.

    (b) It's probably a good idea to start from the last equation. What does that equation tell us if $a^2 + b^2 + c^2 = 1$? Indeed what does it tell us if and only if the original inequality is an equality?

    Because of a small manipulation with the quadratic equality here we can see that we get equality if and only if
    $$a+b+c = 1 \quad \text{and} \quad ab+bc+ca = 0.$$

    Of course the restriction a, b, $c \neq 1$ still applies. Now we want to get to a quadratic in b, that contains no c terms. That will have a discriminant that will have only a's. This discriminant is
    $$\Delta = (1-a)(1+3a).$$

    It's now worth reminding you that the question asks for an infinite number of **rational** numbers, x, y, z. So how is rationality going to come into play at the discriminant stage?

    If we put $a = \frac{k}{m}$, we want $m-k$ and $m+3k$ to be squares. There are a number of ways to do this but putting $m = k^2 - k + 1$ will do.

    I'll leave you to work through the algebra that gives, eventually, $x = -\frac{k}{(k-1)}^2$, $y = k - k^2$, and $z = \frac{(k-1)}{k^2}$.

76. How can you experiment with this one? What values of n is it worth working with? Have you seen anything like this before?

    Let $p \equiv 1 \pmod 8$ be a prime.

Why would the proof start this way? How did the problem setter know to start off the proof this way? What advantages does it have? Is it general enough for what we are trying to do? What would you do next after this start?

The congruence $x^2 \equiv -1 \pmod{p}$ has two solutions between 1 and $p - 1$ inclusive, whose sum is p.

Can you justify this? Where is it leading us?

If n is the smaller one of them then p divides $n^2 + 1$ and $n \leq \frac{(p-1)}{2}$. We will show that $p > 2n + \sqrt{10n}$.

What next then? The square root sign here suggests that we might want to solve a quadratic. What's the obvious quadratic to play with? See what progress you can make before looking at the next piece of text.

Let $n = \frac{(p-1)}{2} - l$, where $l \geq 0$. Then $n^2 \equiv -1 \pmod{p}$ gives $(2l + 1)^2 + 4 \equiv 0 \pmod{p}$.

This needs a step of algebra that shouldn't be too hard. But where to now? What direction can we go from here? What things have we got above that can be useful now? How can we get back to the value of p that we want?

So $(2l + 1)^2 + 4 = rp$ for some $r \geq 0$. As $(2l + 1)^2 \equiv 1 \pmod 8$ and $p \equiv 1 \pmod 8$, then $r \equiv 5 \pmod 8$. This gives $r \geq 5$.

But why is $(2l + 1)^2 \equiv 1 \pmod 8$? What little trick gives us this? You might need to use the fact that, of any two consecutive numbers, one is even.

Now we have that $(2l + 1)^2 + 4 \geq 5p$ and so $l \geq \frac{1}{2}(\sqrt{5p - 4} - 1)$. Let $u = \sqrt{5p - 4}$ for simplicity. Then $n = \frac{(p-1)}{2} - l \leq \frac{1}{2}(p - u)$. All this leads to the quadratic

$$u^2 - 5u - 10n + 4 = 0.$$

For $u \geq 0$, we can solve this quadratic for u in terms of n. You should be able to manipulate the algebra now to get the inequality between p and n that we want. But where does the assumption $p \equiv 1 \pmod 8$ come in? We wanted an infinite number of positive integers n that would have this sort of prime divisor p. How many primes are there that are congruent to 1 mod 8?

77. Play with a few values to see if you can find $f(1)$, $f(2)$, and so on. Note that $f(0)$ is undefined so we shouldn't waste time trying to find it. Is f a simple function? Check out a few simple functions to see if they work. Where to from there? What happens if you experiment with a range of values of p, q, r and s?

Setting $p = q = r = s = 1$ gives $f(1)^2 = f(1)$. Since $f(0)$ is undefined, $f(1) = 1$.

It would be good if we could get an equation involving $f(x)$ that we could solve. What are the chances of doing that? How could we do that?

Let $p = x, q = 1, r = s = \sqrt{x}$.

Some algebra now should lead you to the conclusion that, for every $x > 0$, either $f(x) = x$ or $f(x) = \frac{1}{x}$. You should also be able to see that if $f(x) = x$ or $f(x) = \frac{1}{x}$, then the original condition of the problem is satisfied. The problem now is to make sure that no other function satisfies this condition. But the algebra that led to the conclusion $f(x) = x$ or $f(x) = \frac{1}{x}$ shows that $f(a) = a$ or $f(a) = \frac{1}{a}$. How can we use this to make progress? What other functions might be possible? How about $f(x) = x$ for all $x \leq 3$ and $f(x) = \frac{1}{x}$ for all $x > 3$?

If $f(x)$ is not x for all x, then there exists a positive b such that $f(b) \neq b$. From the work that you did above, this must mean that $f(b) = \frac{1}{b}$. Similarly, there exists a positive c such that $f(c) \neq \frac{1}{c}$, so $f(c) = c$. In the original condition of the problem let $p = b$, $q = c$, and $r = s = \sqrt{bc}$.

Find the value of $f(bc)$ in terms of b and c. Use the fact that $f(bc) = bc$ or $\frac{1}{bc}$ to contradict either $f(b) \neq b$ or $f(c) \neq \frac{1}{c}$. In the process you will have shown that f can only be one of two functions.

78. This is a combinatorial problem and lends itself to experimentation. What is happening here with small values of n and k? Can you see how to do the two types of k-step sequences? Is there any link between the two? How can you relate the numbers N and M? Are M and N finite? Does it matter? Will drawing a diagram help? Have you seen a problem like this before?

To make life easier, let's call any sequence of k switches that leads to the first n lamps being on and the last n lamps being off, an **admissible process**. If that process does not touch the last n lamps call it **restricted**.

First note that in every admissible process, the first n lamps go from off to on and so it is switched an odd number of times. On the other hand, the last n lamps are switched an even number of times as they go from off to off (even if the sequence is restricted). Since k and n have the same parity, it is possible to switch the first n lamps just once and any one particular one of these lamps a further $k - n$ times. This shows that there are restricted processes and so $M > 0$.

I know that this question is too early to ask, but is that last paragraph necessary to the proof? Come back later and check to see where or if it fits in.

Now let P be any restricted process. Take any lamp l with $1 \leq l \leq n$, and suppose it was switched k_l times. Recall that k_l is odd. Now take an even subset of these switches and replace each of the subset by a corresponding number of switches to lamp $n + l$. This choice of even subsets can be done in 2^{k_l-1} ways because this is the number of subsets of even cardinality in a k_l-element set. There are 2^{k-n} ways of combining all of these switches. Their combined action leaves the lamps in the same position as they did in the original process P.

It might be useful to act the sequence of switches out with a small value of n and k. Justify the claim that there are 2^{k-n} ways of combining all of these switches. Do you have a conjecture now for the ratio of N to M? What would you need to do now to prove that conjecture?

We will now show that *every* admissible process Q can be achieved by the method above. But this is not hard because we can transfer the switches of the lamp $l > n$ to the lamp $n - l$ and produce a restricted process.

Is it clear that this restricted process is admissible?

We have therefore set up a $1 : 2^{k-n}$ map between the M restricted admissible processes and the total of N admissible processes. So $N/M = 2^{k-n}$.

But surely there are other ways to go from a restricted process to an admissible process? We could send the switches of l off to a lamp other than $n + l$. Wouldn't this affect the $1 : 2^{k-n}$ map?

80. Draw a diagram. That should at least help you to understand what the problem is asking. Now try it with various triangles to see if you can locate the point of intersection of w_1 and w_2.

To be able to understand the proof of this question you will need to know the results of the following two lemmas and know what homothety is. It would be good if you could prove the lemmas.

Lemma 1. *Given a convex quadrilateral ABCD, suppose that there exists a circle which is inscribed in angle ABC and is tangent to the extensions of the line segments AD and CD. Then $AB + AD = CB + CD$.*

Lemma 2. *The incircle of triangle ABC is tangent to its side AC at P. Let PP' be the diameter of the incircle through P, and let the line BP' intersect AC at Q. Then Q is the point of tangency of the side AC and the excircle on AC.*

Recall also that if the incircle of a triangle touches its side AC at P, then the tangency point Q of the same side and the excircle on AC is the unique point on the line segment AC such that $AP = CQ$.

After all those preliminaries, what can you show on the figure below that is equal and unequal? Are any sets of three points collinear? How does that all help?

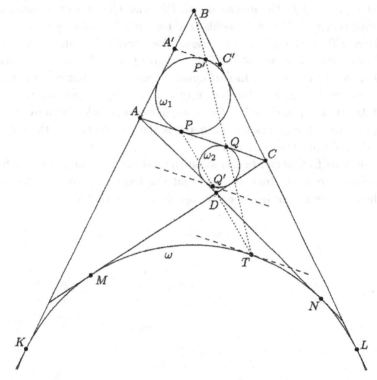

Let ω_1 and ω_2 touch AC at P and Q, respectively. Then

$$AP = \frac{1}{2}(AC + AB - BC) = \frac{1}{2}(CA + CD - AD).$$

Since $AB - BC = CD - AD$ by Lemma 1, then $AP = CQ$. It follows that in $\triangle ABC$ side AC and the excircle on AC are tangential to Q. Likewise, in $\triangle ADC$ the side AC and the excircle at AC are tangent at P.

Why are P and Q distinct points? Would it matter if they weren't?

Let PP' and QQ' be the diameters perpendicular to AC of ω_1 and ω_2, respectively.

What does Lemma 2 show about B, P', Q, and D, Q', P?

Consider the diameter of ω perpendicular to AC and denote by T its endpoint that is closer to AC. The homothety with centre B and ratio BT/BP' takes ω_1 to ω_2. Hence B, P' and T are collinear. Similarly D, Q' and T are collinear.

What homothety gives this last collinearity? What other points can you show are collinear?

We infer that points T, P' and Q are collinear, as well as T, Q' and P. Since $PP' \| QQ'$, the line segments PP' and QQ' are then homothetic with centre T. The same holds true for circles ω_1 and ω_2 because they have PP' and QQ' as diameters. Moreover, it is immediate that T lies on the same side of the line PP' as Q and Q', hence the ratio of homothety is positive. In particular, ω_1 and ω_2 are not congruent.

In conclusion, T is the centre of a homothety of positive ratio that takes circle ω_1 to circle ω_2. This completes the solution because the only point with the mentioned property is the intersection of the common external tangents of ω_1 and ω_2.

81. Solutions to this can be found on the Art of Problem Solving web site referred to in the text. It's not that I'm too lazy to present the proof here rather I'm trying to encourage you to use that site.

Index